"With *Sail to Scale*, Fernandez Guajardo, Jerrehian, and Sabet deliver a masterclass on avoiding the pitfalls that sink most startups. This book is essential reading for anyone who has been charged with steering a young company through the early stages of growth — offering invaluable insights and actionable takeaways."

— Alka Tandan, Chief Financial Officer at Gainsight

"Scaling your company takes more than launching a great product — it also means facing daily unexpected challenges. In *Sail to Scale*, Fernandez Guajardo, Jerrehian, and Sabet give readers clear, actionable advice to avoid common pitfalls and execute successfully through the challenges of startup scaling."

— Ami Vora, CPO at Faire, former VP at Facebook & WhatsApp

"Throughout my wide-ranging career in Silicon Valley, I've worked in, invested in, and advised a multitude of entrepreneurial companies. Some were wildly successful — Tesla, Zoox, and Crunchyroll; others didn't pan out. I developed deep empathy for the entrepreneurs and wished I had a practical guidebook to counsel and assist the founding teams every step of the way. This fine book does just that, arriving at a time when technology is accelerating and the promise and impact of startups is greater than ever. I recommend this great resource to all the courageous entrepreneurs in startups and driving new ventures within larger organizations."

— Laurie Yoler, Partner and Board Member, Playground Global

"If you're serious about navigating the tumultuous journey of building a startup, *Sail to Scale* is an absolute must-read! An essential book for founders or those interested in the lessons of entrepreneurship, unpacking critical and common mistakes made but also the how to avoid them in the first place. A definite winner!"

— Kara Goldin, Founder of Hint Inc.; Author of *WSJ* Best Seller *Undaunted*; Host of The Kara Goldin Show

"I got to see first-hand the mountains that the authors' entrepreneurial spirit could move. We are lucky that they decided to capture their lessons learned in *Sail to Scale*. I highly recommend this book for builders and anyone who wants to be in the driver's seat of innovation."

— Fidji Simo, CEO and Chair at Instacart, Board Member at OpenAI and Shopify

"In *Sail to Scale*, Fernandez Guajardo, Jerrehian, and Sabet distill decades of experience into actionable insights. This book is a game-changer for anyone steering their business through turbulent times."

— Nolan Bushnell, founder Atari and Chuck E. Cheese

"I have worked with entrepreneurs for 25 years as an investor and have seen every mistake in the book — *Sail to Scale* provides wise counsel and a road map I wish had been available for all of my entrepreneurs. It breaks the cycles of business into phases with critical analysis, timing, talent and orientation at each of these waves. This is an absolute must read for all entrepreneurs and investors!"

— Julie Castro Abrams, General Partner, How Women Invest

"Silicon Valley was built on thousands of successes. It was also built on hundreds of thousands of mistakes. Avoiding mistakes — large and small — is just as important to a business' success as its product, strategy, and team. *Sail to Scale* provides a framework for recognizing and avoiding the common errors that can damage or sink a promising startup. It's packed with pointers for the promising entrepreneur, the venture capitalist, or a board member seeking to guide the startup to a successful outcome."

— William B. Elmore, General Partner, Handshake Ventures; Founder, Foundation Capital

"Fantastic stuff. *Sail to Scale* is aimed towards anyone who wants to build an exponential company, is building one now, or has already built and is reflecting upon the journey. I found it to be a great balance of practice and theory, peppered with stories and insights, drawing also extensively upon all three authors' wealth of experience."

– Amit Garg, Co-founder / Managing Partner of Tau Ventures

"*Sail to Scale* is mandatory reading for everyone who wants to build and nurture the entrepreneurial spirit from a place of radical realism. This is not only for those that want to start a business but for those that want to accelerate innovation and total business transformation in well established businesses."

– Antonio Lucio, EVP and Chief Marketing and Corporate Affairs Officer at HP Inc. Executive Fellow at Yale School of Management

"Imagine embarking on a long, arduous, and complicated journey with no roadmap, and little certainty what the next wave will bring. Having now been in seven startup companies, this image is not hypothetical. It's a description of the journey all founders experience. This book changes everything. It not only highlights common mistakes made in these journeys, it's a literal GPS for navigating the challenging, but enormously rewarding experience of breathing life into an idea."

– Richard Mirabile, Ph.D., Founder and CEO, Success Factor Systems, (now SAP SuccessFactors)

"*Sail to Scale* is a treasure trove of practical tips, backed by case studies, on how to avoid a large number of common mistakes in venture building. The mistakes are organized by lifecycle stage, which makes the book easy to navigate. Every founder (and VC!) should read this book."

– Frode Odegard, Founder & CEO, Post-Industrial Institute

"There are plenty of books out there giving inspirational or aspirational advice to entrepreneurs, but very few that give entrepreneurs the nitty-gritty down-to-business guidance they truly need. *Sail to Scale* is one of those rare gems."

– Ellen Taaffe, Clinical Associate Professor, Kellogg School of Management, Northwestern University

SAIL
TO
SCALE

Steer Your Startup Clear of Mistakes from Launch to Exit

MONA SABET

HEATHER JERREHIAN

MARIA FERNANDEZ GUAJARDO

Published by How2Conquer
Atlanta, Georgia
www.how2conquer.com

How2Conquer is an imprint of White Deer Publishing, LLC
www.whitedeerpublishing.net

© 2024 by Maria Fernandez Guajardo, Heather Jerrehian, Mona Sabet

First edition, August 2024
Ebook edition created 2024

Illustrations and cover design by Telia Garner
Edited by Michelle Newcome, Lauren Kelliher, Charlotte Bleau, Emily M. Owens, and Karen Alexander

Library of Congress Cataloging-in-Publication Data is on file at the Library of Congress, Washington, DC.

Hardcover ISBN 978-1-945783-30-2
Paperback ISBN 978-1-945783-36-4
Ebook ISBN 978-1-945783-37-1

For information about special discounts available for bulk purchases, contact How2Conquer special sales www.how2conquer.com/bulk-orders

To our families. Your unwavering support, endless love, and boundless patience have made this journey possible. This book is for you, with all our hearts.

"There are lessons I love you too much to teach you, so you'll need to go out and learn them for yourself."
— *Jimmy Soni*

Contents

Foreword

In the exhilarating but tumultuous journey of start-up life, the path is often circuitous requiring a lot of backtracking, bushwhacking, and moments of feeling lost before finding the success we often read about in the press and from founders around us. As a pre-seed and seed stage venture capitalist and lecturer in the School of Engineering at Stanford for over fifteen years, I've had the privilege of working closely with countless entrepreneurs at the earliest stages of their endeavors. Throughout this time, I've seen up close that what looks like the overnight success of some entrepreneurs is actually a series of profound challenges, lucky breaks, stunning victories, and sleepless nights.

One of the most striking but often unspoken aspects of this journey is how lonely it can be. In a world where nearly every founder I meet appears to be "killing it" from the outside, the reality is actually much more challenging. Particularly in the earliest parts of building a startup, most founders are, by definition, in uncharted territory, challenging conventional wisdom in some way. At Floodgate, we think of entrepreneurs as people who literally bend the curve of the present to help us get to a different tomorrow. But how is that done?

This book is a beacon of light for those who find themselves in this early and solitary stage of building a startup. Written by entrepreneurs who have also operated at scale, it distills the hard earned lessons and insights gained from having personally navigated the twists and turns it takes to build a substantial enterprise. Their stories and experiences are powerful reminders that you are not alone in your struggles and that the challenges you face are ones others have traversed as well. Each story, each example, is infused with lessons that help diagnose your situation and how you might find your way out. From team issues, to when and how to pivot, to transitioning from a jack of all trades to a leader of a scaled organization, and navigating the final stages of getting to an exit, you will find timeless examples of mistakes made and lessons learned.

In reading this myself, I found comfort in knowing that the experiences within some of the companies I worked with were not exceptions to the rule but rather a part of the shared experience of entrepreneurship. Let these stories serve as a reminder that mistakes are not a sign of failure but rather integral parts of truth seeking and learning. I hope that these stories make the journey more tangible, the problems more diagnosable, and the solutions more accessible. They certainly did for me.

—Ann Miura-Ko, Co-Founding Partner at Floodgate

Introduction

The entrepreneur's bookshelf overflows with great books. Some are tales of unicorn startups like Airbnb, Facebook, and Uber, and they ignite dreams of explosive growth and unimaginable success. Others offer step-by-step frameworks for building businesses and products, guiding readers toward achieving "product-market fit," and scaling their ventures in an academic manner. Yet, amid this abundance, we felt a crucial voice was missing: the voice of the entrepreneur in the trenches, navigating the daily challenges and gritty realities faced by the vast majority striving to survive and thrive.

The realization that there was a voice missing in the world's startup library struck us during an offsite in Santa Cruz, a coastal California town by the Pacific. From our base in the Silicon Valley, the heart of entrepreneurship and tech, Santa Cruz offered a refreshing escape just a short drive down Route 17. We'd been discussing our entrepreneurial journeys, beginning with Heather's recent experience as the CEO of Hitch, leading the company through a major pivot and then a successful exit. Her story was a great source of inspiration for writing *Sail to Scale*. As we compared our experiences, moments of "do-or-die" emerged across four potentially destructive waves that nearly all startups face along their voyage: Launch, Pivot, Scale, and Exit. These Four Waves, like ocean swells, could either propel us forward or capsize our ventures. We began to see patterns emerge within these waves, whispering untold stories of success and failure across our combined experience. What struck us most was the dearth of practical, relatable literature dedicated to navigating these crucial and treacherous junctures.

We also realized that, even though the three of us had experience across each of the Four Waves, each of us was an expert on a particular part of the journey.

We're serial operators who have led strategy for some of the most iconic software brands in the market, ranging from Maria's deep product experience at B2B tech startups, Meta, and now at Google; to Mona's corporate development and strategy experience with Cadence Design Systems, UserTesting, and VulcanForms; and to Heather's turnaround and growth experience at multiple tech companies including Hitch and ServiceNow. We've led the strategy for startups, some of which were acquired, others of which didn't make it, and others that are still trying. We've all struggled to bring our businesses through the next perilous wave and to the next growth destination.

Combining our diverse perspectives, we embarked on writing *Sail to Scale*, sharing the best actionable insights on how the decisions you make today impact your journey and your ultimate success. While fantastical tales ignite ambition, ours aims to equip you with the knowledge to identify and overcome the most common mistakes that lurk beneath the surface as you navigate each phase of growth, and to steer your entrepreneurial adventure through the often-uncharted waters of a deadly wave.

Our narrative is woven with real-life startup stories, not just abstract theories. Some stories, with the founders' consent, are presented in detail, offering company-specific contexts and lessons learned. Others, respecting personal journeys and privacy, remain anonymous. Failure can be a sensitive topic, and some stories are yet to be told publicly, let alone by us.

This book aims to be the companion to entrepreneurs at large. Whether you're launching a new company, pioneering a product within a large corporation, or simply curious about navigating the entrepreneurial seas, this is your guide. It's for you, the founder, the executive, the early employee, or anyone passionate about bringing ideas to life who wants to know what lies beneath the treacherous waves of entrepreneurship.

Read *Sail to Scale* in its entirety or delve into specific chapters relevant to your current wave. We recommend the full voyage.

Knowing what lies ahead will prepare you to see the signs as soon as they appear on the horizon.

You might also notice that we — three women — are the co-authors of this business book. Though uncommon, we want to make it clear that while our individual experiences are shaped by our identities, this book transcends gender. This is the only mention of our gender within the book, not because we want to downplay it (we are strong, public leaders and advocates for women in tech), but because the focus here is on the universal journey of entrepreneurship.

Aileen Lee of Cowboy Ventures coined the business term "unicorn" back in 2013 to describe a company valued at a billion dollars or more. A decade later, a startup still only has a 0.00006% chance of becoming a unicorn. Perhaps if the industry talked more about the major inflection points in a startup's journey — where things often go seriously wrong — and focused on advice about how to anticipate them, meet the challenge head on, and ride the wave to and through your next milestone, perhaps then we could create more unicorns. Or better yet, we could build a more robust business ecosystem where more companies can succeed.

So, don your life jacket, gather your crew, and prepare to chart your course. *Sail to Scale* is your compass, guiding you towards entrepreneurial success, one wave at a time.

The First Wave

Launching Your Ship

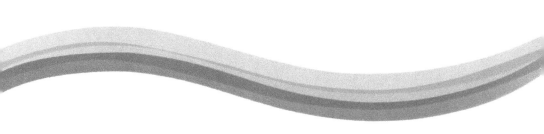

As an entrepreneur, there are many things you must do to launch a new startup. You have to figure everything out in a rush of excitement, with insufficient resources and a massive learning curve — funding, deciding what to build, operations, payroll. Search the web and you'll see more articles on startups than you can consume, with headings like "How to Launch a New Business: Three Approaches that Work," or "Launching a Business in a New Industry? 15 Important Things to Do First," or "11 Steps to Building a Successful Business." These articles are superficial. Nothing of depth can be written about the entire Launch stage in two thousand words or less. It would take an entire library to cover all the aspects of this intoxicating and fun phase.

This section (and this book) isn't a comprehensive outline of everything you'll need to do to get your new business off the ground and running. It's a targeted look at the most common mistakes made when launching a new business or product, and it details the nuances you need to understand to steer your company around or through those mistakes.

As you navigate the Launch Wave, you'll make important decisions that anchor you for the future — what vertical you're pursuing, what your target audience should be, how you're going to make money, your technical stack, who you're hiring, and many more. And while we're advocates of learning and iterating to reach success, the closer you get to success during the Launch Wave, the easier it will be to conquer the Scale Wave.

There are endless mistakes you can make during any of the Four Waves. But the five mistakes we focus on in this section are ones that we've seen over and over in our collective experience, across both startups and new ventures inside established companies alike.

Mistake 1: Being Too Focused on the Short Term is about starting your new venture with an overly narrow focus that limits your ability to develop a thriving, large-scale business. We explore how overcoming this mistake requires you to become a futurist, a strategist, and a business-savvy professional.

Mistake 2: Building the Wrong Minimum Viable Team happens when you assemble an ineffective initial team, which often results from not identifying and recruiting the necessary skills. Mistake 2 will talk about the attributes your core team should possess and introduce the concept of "archetypes" — especially the "entrepreneur archetype."

Mistake 3: Being in Love with the Solution occurs when you're too enamored with the solution you've built. Mistake 3 will consider how letting go of your brainchild is difficult, as it's closely tied to your ego.

Mistake 4: Building the Wrong Minimum Viable Product is about not striking the right balance when building your MVP. You may have too many features or not enough, and you can waste precious cycles trying to figure out why you're stuck in muddy waters and not moving forward.

Mistake 5: Scaling Too Soon explores how the consequences of this mistake go beyond just having to pay for unnecessary salaries. We conclude with the very tempting desire — and often a mistake — of scaling too soon.

Before we venture into the specifics of these five mistakes, let's start with a story that illustrates the journey one startup took as they navigated their Launch Wave.

Story:
UserTesting Your Way to Launch

Dave Garr is a website optimization expert. Early in his career, he managed some of the most well-known websites, including Apple and HP Shopping. As he developed his expertise, he realized the path to successful websites was to show the designs to real people as the site design developed, not just after it was completed and launched.

At that time, the tools and solutions you think of today to help you optimize websites didn't exist. Dave maintained a group of a dozen people he knew who were open to providing him with feedback on the design iterations. Sometimes he would hire a recruiting agency to find people he didn't know to physically come to his office, so he could watch people use his website while thinking aloud. This approach was time-consuming and expensive. The recruiting agency took a long time to find the participants, and then some wouldn't show up, and others who did show up didn't provide much useful feedback. Dave wished he could have quick access to users who were observant and could continuously verbalize their thoughts while using his website.

Darrell Benatar had previously been the founder of Surprise.com, which collected great gift ideas from users and linked to online merchants. Darrell hired Dave to design the website, and through a lot of user testing, Surprise.com became hugely popular and one of *Time* magazine's "50 Best Websites." Dave and Darrell became friends.

One day in 2006, Dave and Darrell met up for lunch at the Sunnyvale Hobee's, a family restaurant in the heart of Silicon Valley, charmingly known both for their blueberry coffee cake and as the place where entrepreneurs and businesspeople talk deals over food. Dave told Darrell about his pain point and his "magic wand" dream. Even as early as 2007, Dave believed that 99 percent of all websites were being built and run without ever having real users give feedback. Dave's approach to website optimization made

Darrell realize the struggles he had with websites and apps were a design problem, not a user problem.

A pain point keenly experienced — in a space without a solution yet — and an idea for a magic wand to fix it? Those were the key ingredients of a new business opportunity, and both knew it.

The first big challenge with the "magic wand" was how to source a trove of people willing to participate as users. To meet this challenge, Dave and Darrell decided to build an MVP under the company name "UserTesting." The first version of their MVP would allow anyone to sign up to be a reviewer online and get paid a small amount ($10) for every recording session they completed. Of course, that meant UserTesting had to find customers willing to pay people to give feedback on their websites. And they knew website owners and designers would want to hear from the specific people they were targeting to use their website, not just anyone who wanted to make a quick $10. So they started by asking the participants who signed up to identify themselves by gender and age range — just enough to give customers some comfort that they weren't talking to a 70-year-old woman about whether she liked a new basketball shoe campaign.

On the customer side, the UserTesting MVP was just as straightforward. The offering was a "$49 per test" self-service offering. The customer could set up a test with a simple form that collected some basic information: a link to the website to be reviewed, a description of what they wanted feedback on, demographics and a description of their target audience, and credit card information. And in return, the customer would get back a video of a real person giving their feedback and opinions on the website they'd reviewed as a user.

Sound too simple for an MVP? Well, remember this was back in 2007. The world (actually, mostly the US) was getting introduced to its very first iPhone. There was no such thing as an app store yet. Skype was all the rage.

Also, it *was* simple — almost too simple. The co-founders had a trickle of individuals signing up to take tests, but not enough to support a lot of tests if customers started pouring in. Which they didn't, at first. Because no one knew you could get this kind of testing (what experts would call "qualitative usability testing") over the internet. So no one was searching online for it.

To solve their discoverability problem, the founders started writing to user experience (UX) thought leaders and online business bloggers to tell them about this new SaaS (software as a service)-based usability testing product. A few researchers and thought leaders in the UX space responded quite negatively; they insisted usability testing needed to be moderated by experts, and testers needed to be thoroughly vetted. Most disruptive technologies threaten the status quo.

But the response wasn't universally negative — bloggers loved the idea. It gave them something easy to write about. UserTesting started finding themselves in articles: "Ten Ways to Make Money from Your Couch." Number one? Sign up for UserTesting.

These articles drove interest in signing up to be a UserTesting reviewer. They also helped raise awareness that there was another way to easily and inexpensively understand how usable a business' website was. Darrell says he still has screenshots saved of comments on Twitter (now X) saying: "This is the best money I ever spent!"

The company started growing through word-of-mouth and managed to achieve $1.5 million in sales with this very lean MVP. And their customers loved them. Was all this customer love proof they had found product-market fit? Some would say yes. And perhaps, in the narrowest sense, they had. But it still wasn't a viable business. While UserTesting was seeing growth in first-time customers, few were translating to repeat users. Customer retention was a problem because website owners found so many problems to fix by running just a small number of tests, they would spend many months fixing those problems before wanting to run more tests. And if you can't retain your first customers, you haven't proven you have a viable business opportunity. Darrell recalls that, about this time, a key advisor told him, "Your business is no longer of interest to professional investors. Either cut your losses now or change your business model."

So the founders took one more step in their business evolution. They added an option to buy a subscription with "unlimited" usage. Customers who purchased this could run as many website tests as they wanted for the term of the subscription. This relieved their customers from worrying about having to pay every single time they wanted to run a test and allowed a behavior of repeat testing to develop.

Around the time the founders were launching this new pricing model, one of their customers called them to say they had some leftover budget to spend before the end of the year, and to ask if UserTesting had an offering they could get their entire team to use. Being a keen entrepreneur, Darrell said, well, in fact they did. And they created their first enterprise subscription.

At the time, UserTesting had a fair number of users that were part of teams at large companies. They started aggressively offering their new enterprise model to these users and got immediate traction. Darrell recalls the company spending a year just upselling subscriptions to existing customers. The product still wasn't designed for an enterprise. The enterprise might not have liked that, but the enterprise users loved it.

There's a lot more to the UserTesting story — about a decade more actually. For the purposes of our Launch Wave, we'll stop here, except to say that UserTesting was acquired in 2021 for $1.3 billion.

Mistake 1: Being Too Focused on the Short Term

It's surprising how many startups, departments, and even entire companies don't have a long-term vision that can support a thriving, large scale business. Planning how to make your product or services valuable and competitive over the long run is crucial — it can't be just an afterthought. You might neglect to invest in long-term thinking because it seems daunting, subject to change, intimidating, and not immediately urgent; it's easier to focus on the smaller wave immediately in front of you. Many entrepreneurs, especially those from a business or marketing background, believe they've created a long-term strategy when all they've really communicated was a 10,000-foot vision — almost an ad campaign. Others, especially those from an engineering background, are often focused on solving immediate problems, finding the idea of business strategy and future planning as appealing as doing 500 burpees.

Startups evolve through iterations. There's no expectation that you'll get everything right the first time, and you won't. So why do you need a long-term plan? Because you don't want to put yourself in a corner. You'll need to make "millions" of decisions every day, and among them, several will be what Jeff Bezos is credited with calling "Type 1 decisions" — big-stakes decisions. These are nearly irreversible, requiring too much effort or cost to change later. Without a long-term plan to use as a compass in evaluating these decisions, you might later find yourself in a difficult situation needing to quickly bail water out of your sinking boat.

Some examples of how this shortsightedness can play out in the future:

✖ You didn't plan for an audience or market that's big enough to support the growth of your company, and now you can't find funding that matches your needs. You run out of cash.

✖ You didn't anticipate how the competition would react, you didn't protect your advantage, and now more established players offer the same product with less friction. You can't find funding, or a mergers and acquisitions (M&A) opportunity, and you run out of cash.

✖ You didn't anticipate the features you would need, and the technology stack you chose doesn't allow for those features to be built easily. You run into huge delays building them, and you run out of cash.

It's good to work with a roadmap that looks at each quarter or year and focuses on gradual progress toward building a great product that people adopt. However, this needs to be based on a strategic, bold, forward-looking plan. And both need to be developed and evolved at the same time.

Practice Long-Term Thinking

Long-term thinking isn't a superpower reserved for the chosen. It's not innate; it can be learned. Maria had to learn this early in her career. After working as a silicon engineer, she became a program manager at Texas Instruments, where she led teams building application processors for smartphones. Maria ensured the work was completed in the correct order, on time, and according to the specifications, while solving any daily problems that arose, often within a tight schedule. Her role was crucial, as her commitments were expected to generate billions in revenue, yet it focused solely on execution in the near future.

After several years of this type of work, Maria's curiosity led her to explore the early definition phases of these products. Given that complex semiconductors typically took three to five years to build, she knew someone had to be planning for the smartphone's future computing needs. This work sounded both exciting and intimidating, prompting Maria to join a team that looked further ahead, one that was closer to the business, the customers, and the users. As she transitioned into the business development team, she feared her execution-focused background might not have aligned with the strategic and business skills required.

Fortunately, Maria had a mentor who encouraged a growth mindset. With deliberate practice, a significant portion of which she undertook in her own time, this approach became second nature to her. Months later, Maria was successfully integrated into the business development team, discussing future trends, business, and strategy with customers and her internal teams. A few years later she moved to product management, leading startups and innovative teams within large companies to develop the products of the future, and long-term thinking became part of her day-to-day.

To successfully become a startup long-term thinker, there are three skills you need to develop:

BECOME FUTURIST

RAISE BUSINESS ACUMEN

THINK STRATEGICALLY

› Become a futurist. To find the biggest opportunities, you need to anticipate changes in trends, technology, and user behavior. The key is to anticipate, and for that, you need to think about the future and start living in it.

› Raise your business acumen. You need to think about how you'll reach profitability earlier on than most advice suggests. You don't have a business without profit.

› Think strategically. Having a powerful vision and lofty goals isn't enough; you need to know how you're going to get there with the resources you have.

Become a Futurist

A startup is rooted in innovation. Something that someone wants is being created or offered through a business model in a new way. The biggest opportunities lie in insights you act upon before they become evident to someone else. Nobody can predict the future, and we're not asking you to invest in a crystal ball. But those who are able to understand the future trends, implications, and second-order effects even just 10 percent better than competitors or incumbents have a leg up.

> The biggest opportunities lie in insights you act upon before they become evident to someone else.

These insights appear when there is change. New business initiatives miss the biggest opportunities when they fail to anticipate changes in trends, technology, user behavior, and so much more. But there is a framework to help you know where you need to pay attention to changes.

Building a new product, service, or business happens at the intersection of three areas that are constantly evolving and creating insights for new business opportunities: customer problems to solve, technology to solve it, and business models to generate value for the business.

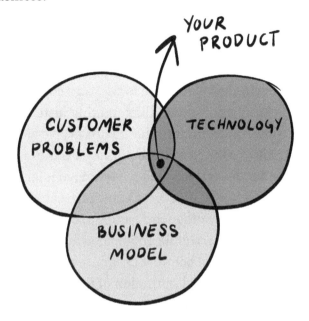

1. Customer problems: The challenges people and companies face change all the time, creating opportunities for new companies to solve them. For example, in 2024, hybrid work and a return to the office created opportunities to solve new problems. Companies needed to manage their employees, offices, and tools to live in this new world, and the startups that anticipated these trends the earliest had a more time to figure out how to solve these emerging problems for their customers.

2. Technology: The technological landscape constantly evolves, sometimes incredibly fast, enabling a whole new slate of tools that can be used to solve problems or make existing solutions better. Recently, the rapid advances in generative AI has fueled startups that automate the creation of everything including personalized marketing campaigns, new video content, and computer code. For example, tools

like MidJourney are enabling artists, designers, and a broad set of professionals to create stunning visual artworks and prototypes rapidly in ways that were not possible before.

3. Business model: The new product or service must generate revenue and eventually profits from the value it creates. Innovating in the value chain is also very common — creating bundles or unbundling, renting instead of buying, etc. Business models and value chains evolve and create new opportunities. One example in this area is the change in user behavior Netflix created with streaming media. This change drove people to cut their more expensive cable subscriptions and inspired many other companies to create similar streaming services. Netflix anticipated and acted upon a new business model before it became evident to others. Now it seems we've reached saturation of this business model, and we're on the brink of a new model that better serves the over-subscribed viewers.

> Value chain is the "chain" of steps or activities that go into the creation of a product in order to create value.

Not all three areas need to change at the same time. Addressing only one may be sufficient to create a big enough future business opportunity. An old problem is now solvable with new technology. Or a new problem emerges that can be solved with existing technology. Or an existing solution can be offered with a different business model. The point for an entrepreneur is to find a winning combination before it becomes obvious to everybody else.

Let's look back at our Launch Wave's original story on User-Testing. Founders Dave and Darrell exemplify how to think like a futurist in the early stages of your startup. They identified the changing trends across customer pain points, technology, and business models that created the opportunity for a company like UserTesting to exist and thrive for the long run.

Customer Problems	Website designers had no efficient way to optimize their websites for the best user experience. They had to spend a great deal of time and money on research recruiting agencies, or they would just launch their sites and cross their fingers, hoping it met their users' needs.
Technology	Back in 2007, the iPhone was just launching, making it easier for individuals to record their feedback on any website served up to them through an app like the one UserTesting would soon build.
	In addition, cloud-based applications were exploding, making it significantly easier to launch an interactive website. PayPal launched in 1999, making it simple to get online payments. Shopify launched in 2006, making it simple to put up an online store. Stripe launched in 2011, making it simple to manage all your business payments online. What was going to differentiate one website from another? The UserTesting founders saw two things before it became evident to others. First, with cloud services, it would be much easier to stand up new web services and products, creating a big market. Second, with the democratization of website functionality, user experience would be the differentiation among websites and would become a problem that site owners would pay to address.
Business Model	The founders initially set up a pay-as-you-go online payment model for customers, so customers could get feedback with minimum friction and without having to commit to a long-term contract just to optimize their website.

So how do you become a futurist? How do you find an idea and a strategy that's worth pursuing over the next few years before it becomes evident? It's totally a learnable skill and a lot of fun once you master it. At the core, there are two main things you need to do. First, you need to be informed and in the know, and second, you need to anticipate changes.

Become a futurist by finding an idea and a strategy that's worth pursuing over the next few years before it becomes evident.

Being in the Know

Being in the know means being informed about society and trends, technology, and the business environment. Being an avid reader or listener, actively seeking information, and living in the forefront of trends demands hard work, and it requires a significant amount of time. But there's no need to be discouraged. More and more tools are coming up that summarize information from massive amounts of sources, tailored to your needs. And as you keep up with information and stay more and more up to date, over time it becomes easier since you'll only need to pay attention to the new information that came up that day or week.

Being in the know in society

To imagine what products and services will be needed in the future, start by learning about and questioning the world around you. How is society and human behavior changing and adapting over time? What are the current trends, and which ones can we anticipate? For more mainstream analysis, read publications like *The Atlantic* or the culture and lifestyle sections of *The Guardian*. Or listen to society podcasts like *Radiolab*. And of course, most nonfiction sections at your local bookstore or library will have a variety of topics that may illuminate how society and human behaviors are changing.

If your business sells directly to consumers or builds a product for youth, trends are evolving rapidly, and you have to be a lot closer to where they're developing. This may require you to spend time on the latest social media app or on the ever-enjoyable exercise of asking teens for their opinions and input. It's equally important to know which human behaviors are not changing. Some things are so ingrained in us that they persist through trends, like the need for people to connect with other people. A good primer to read about these fundamental people "truths" is *100 Things Every Designer Needs to Know About People* by Susan Weinschenk.

Being in the know in technology

Being in the know requires keeping up with the latest tech trends — everything from tech stack choices, research progress, developer chatter, etc. Even if you don't consider your business a technology business, technology is likely a critical enabling function for your business. And companies that embrace technology are the ones that are more likely to shape the world. A big source of news to "be in the technology know" is *Hacker News*. And, of course, all sorts of podcasts, YouTube channels, social media accounts, and LinkedIn posts are accessible, depending on your technology platform of choice. Attending technology conferences can be useful as well for any topic you want to explore in depth, to understand the timelines and risks for practical implementations. Research conferences are a favorite. They're hard to digest, but researchers are living in the future, so hanging out with them will give you advanced insights.

Being in the know in business

For keeping up with the never-ending founding rounds, business models that are resonating with the market and investors, and business news in general, a good resource to check daily is TechMeme.com. It's a good aggregator of tech news, but it also shares tech influencers' responses to the news, which provides added color. Another great resource is the quarterly earnings reports and the annual shareholder letters from publicly traded companies. Although the financials of big companies may seem far away from your startup, they usually address how future macroeconomics will impact their businesses (and possibly the future of your startup), and help you learn how to think outside-in (see **Mistake 16: You're Inside-Out**).

Anticipate Changes

The most important part of becoming a futurist is not just gathering information by, for example, attending conferences and listening to a podcast. That's only the beginning. You need to ask yourself, "So what? What would the consequences of this change be?" That's how you'll start developing intuition anticipating the effects of those changes. In addition to reading, listening, and researching about the future, you must intentionally think about the future — which may sound obvious, but it requires active practice.

Adopt an inquisitive state of mind

Whenever you see the news, a discovery, or a trend coming up, force yourself to think about the consequences of that event across a large range of topics. For example, people are having kids later and later in life. What does this knowledge suggest for how people's behavior will be impacted tomorrow? What about in five years? And in ten years? What are the second- or third-order consequences? What will be the impact on the economy? What products or services will be needed in that new world?

In UserTesting's case, the consequence of cloud-based business applications was that websites, and eventually mobile apps, would no longer be able to easily differentiate based on functionality — they were all migrating to using third-party backend software instead of building their own. The impact was that website visitors and app users started deciding which website and app to use based on the experience of using the app rather than the backend functionality it offered.

Discover and discuss ideas

Debating ideas, merits, and potential issues — and questioning your own judgment — will make your ideas and ability to anticipate change stronger. There are forums dedicated to thinking and talking about the future. For example, in San Francisco, The Long Now Foundation offers talks — in person and livestreamed — that explore futuristic ideas and their impact on society. One of their principles is: "To get to a future you want to live in, first you have to be able to imagine it." Another place to find thought-provoking conversation starters is the Reddit forum "Futurology."

Or you can bring your idea generation and discussion in-house. A startup we talked to created a chat channel called "Bad Ideas," so employees felt comfortable sharing and discussing any idea without a preconceived expectation that it had to be a good one. You can also gather a group of future-minded colleagues or partners and set up time to talk about the more distant future. When Maria was working at Meta as a product manager in the Oculus virtual reality division, people were thinking and talking about the future all the time — in fact, there were too many ideas. Virtual reality was such a powerful technology that the opportunities seemed endless. During that time, she partnered with a future-minded product manager on a monthly "idea garden" session, in which they let their imaginations and creativity run wild. They would pitch ideas to each other, passionately debate, and discard most of them without bothering the rest of the teams that needed to work, somehow peacefully, on a well-defined deliverable.

Live It

Our final piece of advice about becoming a futurist: use and adopt as many new products yourself as soon as you can. In the curve of innovation adoption, the first people who use a new technology or a product are the innovators, who represent about 2.5 percent of the population.[1] Being on that side of the curve means using products that are half baked, that require a lot of commitment to making them work in your life. And if you're living in the future, you get used to connecting the dots and seeing the possibilities versus complaining about something not working perfectly or a feature not being available yet.

Raise Your Business Acumen

Raising your business acumen is about learning how to create a financially viable product and business in the long run, like a CEO of an established company, which is what you would expect your company to become. Developing this knowledge during your Launch Wave is equivalent of "dress for the job you want." If you don't already understand them, start by learning basic business financial concepts (there are endless mini-MBA classes you can take), and then dive into the specifics of your chosen industry. Learn about the specific metrics your industry measures companies against. For example, SaaS companies are measured by a variety of SaaS-specific financial metrics. If you happen to be in the SaaS industry, we've added a primer on some key metrics in the Sidebar at the end of this chapter.

Understanding not only what they are, but also what the industry averages are, and when you should and should not be worried if you fall below any of those averages are critical elements of business acumen for being able to assess how your startup is measuring up against what you targeted your big future vision to be.

Because you need to think about profitability from the beginning, you need to learn how value is created and how money is being generated and distributed end-to-end in your industry. Learn about the value chain, incentives, competition, and the

---

[1] Rogers, "Adopter categorization," 247.

forces and factors that shape your specific industry, like regulation or buyer's power. A great way to start is by reviewing the earnings reports of leading potential customers, incumbents, or suppliers, or by reading industry-specific business publications. Networking and engaging with industry professionals, entrepreneurial or established, will also raise your business acumen, particularly if you skip the chats about your ski trips and prioritize asking business or industry questions.

Think Strategically

The term "strategic thinking" has often been misunderstood and misapplied. "Strategic" has been stripped of its meaning and floats around as an adjective for everything and anything: strategic objectives, strategic goals, strategic vision, strategic planning, and so on. The situation has gotten so bad that we even considered renaming this section, or just getting rid of it. But we're on the side of reclaiming the real meaning of strategy — and having one — because it's needed. So here we are. Let's start with what strategy is not.

✗ Strategy is **not** having a vision. You need a vision of the future to be able to shoot for something. Envision what your users, what the world, would look like if you're successful in your endeavor. However, having a vision isn't the same as having a strategy.

✗ Strategy is **not** having goals. Yes, you need goals, and hopefully after having raised your business acumen, you have goals that help you win the market, rather than barely existing in it. But having goals isn't the same as having a strategy.

So, what is strategy? Strategy is *how* you achieve your lovely vision of the future and smash winning goals. It's the real deal.

At a minimum, strategy should:

✓ Have phases

✓ Iterate

✓ Be rooted in strengths

✓ Account for the rest of the world

Strategy Phases

Anything worth doing is going to take some time, and with a startup, you won't have a lot of means to build your entire vision all at once, so you need to sequence things. And those phases usually either enable or build on each other and provide proof points that the path taken is the right one along the way.

A classic example of these phases is the Tesla strategy. Even if the founders had the ambition to make the electronic vehicle (EV) pervasive, they knew they couldn't start building mass-market cars like the Model 3 right away. They didn't have the scaled technology they needed to produce cars to satisfy the high volumes required by a mass-market car, and it would've required massive amounts of investments in a market that hadn't been successful up to that point. Their phased strategy was to start with a high-end model, the Roadster — a low-volume car used as a first phase of their strategy to help finance the development of many of the foundational pieces they needed for large scale production. A slightly less high-end model like the S would follow and so forth. That allowed them to build more and more scalable technology for cars and manufacturing, fund operations at a bigger scale, and turn around market perceptions on EV.

Strategy Rooted in Strengths

Good strategies reflect a deep understanding of the problems they're trying to solve and a strong diagnosis of your competitiveness. They're rooted in the company's strengths. Your company's strength may be key talent, technology differentiation, access to proprietary data sources, affinity access to potential customers, or something else entirely. Your strengths form the basis of your competitive advantage. What is it you're doing better than the rest? A good diagnosis of the situation and your ability to address it are key ingredients for a good strategy.

Accounting for the Rest of the World

Your startup isn't going to operate in isolation — that would be too easy. There's competition, there are changes in the regulatory environment, and there are changes in behavior and technology. You can't anticipate all these things, but some you can. Such as your competition — if your strategy works and you start winning, how does your competition react? Do you have a plan, an advantage that will allow you to keep winning? For example, if your startup depends on a big company not launching one easy feature in their platform, your shelf life might not be long.

Many leaders we've worked with love to create SWOT analyses as a framework to systematically account for internal and external influences to your company or product.

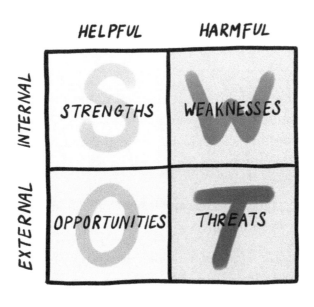

SWOT stands for Strengths, Weaknesses, Opportunities, and Threats. And we like SWOTs too. But many leaders don't know why they're creating their SWOTs, or what to do with them after they've spent an entire offsite with their leadership team filling out the four quadrants of a SWOT in excruciating detail. A SWOT is a strategic tool for accounting for the rest of the world. But it's wasted time if it isn't updated regularly — quarterly given the speed of change in today's business world. And it's even more of a waste of resources if it doesn't end up driving revisions to your strategy.

Iterative Strategy

Strategy isn't a "set it and forget it" type of thing. It's a process, not a fixed destination. You'll iterate on your strategy, just like when developing your product. Your strategy relies on assumptions that you're making about the product, the market, and the competition. And as you go along and learn more, you'll need to readjust your hypothesis and your plan. For rapidly moving markets (e.g. consumer internet), you should review progress against your assumptions every quarter; for slower moving markets (e.g. automotive), every six months may be enough.

> Strategy isn't a "set it and forget it" type of thing. It's a process, not a fixed destination.

Practicing long-term thinking and doing the hard work of becoming a futurist, raising your business acumen, and developing your ability to think strategically can feel daunting. But just like designing, accounting, or even golfing, this too is a skill that can be learned.

If you're a fan of models and frameworks, then here are a few books we like on strategy:

> ❯ *7 Powers: The Foundations of Business Strategy* by Hamilton Helmer

> *Good Strategy/Bad Strategy: The Difference and Why It Matters* by Richard Rumelt

> *Blue Ocean Strategy: How to Create Uncontested Market Space and Make the Competition Irrelevant* by W. Chan Kim and Renée Mauborgne

You can also learn a lot through stories and examples as they develop in the world. The website Stratechery.com is an excellent source of reflection on big players' strategic moves.

As with everything you learn, you must practice strategy before you need it. Even if you don't get a lot of opportunities to practice thinking strategically in your current role, there are ways to start practicing today. Pick a few of your favorite companies and predict their business moves in the face of change. Then observe, reflect, and track what you got right and what you didn't. Having a set of theoretical frameworks and practicing mapping the world of business to them will help you start thinking strategically much more easily.

Key Takeaways

---

1 In your enthusiasm for getting off the ground, it's easy to focus too much on the short-term. Without a long-term vision that anticipates changes in trends, technology, and user behavior, your new initiative is likely to miss the big opportunities that create the foundation for a scalable business.

---

2 Become a futurist by always remaining curious. Stay informed by reading, discussing, and questioning the things you see changing around you. And then ask yourself how you think those changes are going to play themselves out in society and in business.

---

3 Raise your business acumen and deeply understand the business you're in. Profitability can't be (too much of) an afterthought.

---

4 Strategy is not about having vision or goals. It's about how to achieve them in phases over time. Good strategy is rooted in current reality but can take on many iterations over time. It's a process, not a fixed destination.

Sidebar:
SaaS Metrics

In the pre-SaaS world, most software vendors had to charge an upfront one-time license fee to account for the fact that the customer would have access to the software forever. The fee was generally large to account for that fact, regardless of whether the customer actually used the software or not once they had it in their possession. These one-time fees made for a lumpy revenue profile and made forecasting difficult.

Enter the SaaS business model loved by CEOs, Chief Revenue Officers, and investors for its recurring subscription fees for monthly (or yearly) access to software, enabled by the fact that accessing software through the cloud (instead of downloading it) allows the vendor to fix issues and deploy them, provide new functionality and upgrades, and turn off access if needed at the end of a paid period. A significant side benefit is that customers don't have to shell out millions of dollars at the start of a license, since they aren't paying for a lifetime of access, but just a month or a year at a time. As long as the product delivers value, the customer keeps renewing its license and paying for continued access, creating a more predictable revenue stream than the old, lumpy software download model. Investors *love* predictable revenue streams. And so, investors have loved SaaS business models for a long time now.

The Value of SaaS

But all that predictability and love are based on two big assumptions:

1. That the customers will actually continue paying for access with each new subscription period; and
2. (Relatedly) that the amount it costs to get a new customer to become a subscriber (sales and marketing costs) is less than the amount you'll make from the customer over time before the customer stops renewing — because at some point, even the best products get replaced

And thus, the first set of SaaS metrics were born — or at least brought into the mainstream.

First, since a customer could stop paying for access to the software before the next subscription period (monthly or annually), the advantage of recurring revenue could disappear quite quickly. Customers who have terminated their subscriptions have "churned."

Churn is the total number of customers or subscribers who have stopped using a service or product over a specific period.

Churn rate refers to the percentage of customers who cancel their subscriptions or stop using the software service within a given period.

Second, a customer might renew their subscription for multiple periods before they stop paying for or using the product. However, if the total fees you received from that customer during the entire period they were your customer — called the Customer Lifetime Value (CLV or LTV) — is still less than the total amount you paid in sales and marketing expenses to find and acquire that customer in the first place — called the Customer Acquisition Cost (CAC) — then you're losing money on that customer — not a strong business proposition.

CLV/CAC is the measure of the return on investment in acquiring new customers by comparing the value a customer brings over their lifetime (CLV) to the cost of acquiring that customer (CAC).

Customer Lifetime Value (CLV) is the projected revenue a company can expect from a single customer during their entire customer relationship.

Customer Acquisition Cost (CAC) is the total cost incurred to acquire a new customer, including marketing expenses, sales commissions, and other related costs.

Churn rate and CLV/CAC. These are two key metrics that drive a decision about whether your SaaS subscription business model is in fact preferable to the "perpetual" software models of old.

Of course, these aren't the only metrics that are important in evaluating the health and performance of a SaaS company. And early in a startup's lifecycle, you won't have enough customers or history to measure even these two metrics with any reasonable level of confidence. But for now, let's dive deeper into each of these metrics and then introduce the rest of the SaaS metric slate.

Churn Rate

Churn rate is a measure of the value you deliver to the customer, because a customer that has the option of not renewing access to your software for the next period will eventually take that option if they aren't getting sufficient value from your software.

> Churn rate = Number of customers that canceled a subscription (or in the case of freemium models, that no longer engaged with the product) during a period ÷ Total customers at the start of that period

As your business grows, hopefully so does your number of customers, but often so does your churn rate. That's why we usually don't consider a churn metric for a very early-stage company that hasn't been in market for more than two years.

For a mid-to-later stage enterprise B2B SaaS company, a good churn rate is around 5 percent.[2] If you're selling into the SMB (small- and medium-sized business) market, a good churn rate might be more like 10 percent. In the second quarter of 2024, Netflix (B2C) had a very low churn rate of 2.5 percent.[3]

If your churn rate is high early in your company's life cycle, this could be an indication that you need to pivot. It may be an indication that the customers you're selling to aren't your ideal customer profile.

Some people measure churn in dollars rather than number of customers.

---

[2] Murphy, "What is a Good SaaS Churn Rate."
3 Friedman, "Netflix Boasts," MediaPost.

Dollar churn rate = Revenue lost from canceled subscriptions during a period ÷ Subscriptions revenue at the start of the period

Customer Churn Rate and Dollar Churn Rate will tell you two different things. Here's an example:

Best B2B SaaS Inc. (B3S) offers a SaaS solution for companies wanting to build community among their customers. They offer a monthly subscription model based on the number of community members the company has.

In their last quarter, their customer churn rate looked like this:

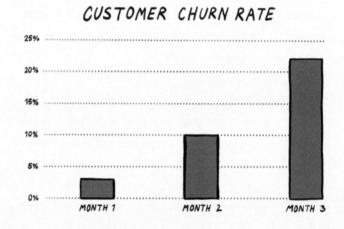

In the same quarter, their dollar churn rate looked like this:

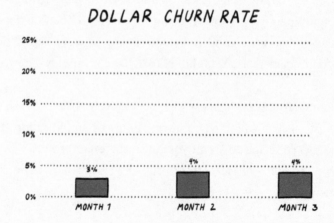

The lower the churn, the better. These two charts appear to tell two different stories. B3S is losing more and more customers each month, but that doesn't seem to be affecting their dollar churn rate much. If you looked at customer churn on its own, you would get very worried. If you looked at dollar churn rate on its own, you might be very happy. When you look at them together, you realize you need to look deeper.

Shannon Power, CFO of Scope AR, explains it this way:

"In this scenario, you are obviously losing the customers who are spending less with you, since your Dollar Churn isn't increasing like your Customer Churn is. Initially, you may think that's a great thing — get rid of your lower paying customers and keep your higher paying ones.

"But looking deeper into your churn data might reveal other insights. For example, you may find out that the customers churning are mostly from a particular customer segment. If that customer segment represents a much larger TAM than the customers in the segment you are retaining, you might have short term growth but longer-term pain.

"You may want to combine your churn data with your sales pipeline and conversion data. If you've recently seen a large increase in your lead-to-close conversion rates, your sales team might be bringing on new customers without qualifying them properly.

"Or you may want to consider your churn data in conjunction with any recent changes in your business strategy. Perhaps you just made a conscious decision to try to sell more into the SMB space to capture market share as companies grow. These churn numbers may be explained by that change in strategy and also inform you on how that strategy is doing."

A deeper analysis of churn rate helps companies identify patterns and trends in their customers' behavior, which should feed into your overall operational strategy. While it's easy to define what churn is, you can see that understanding the reasons and impact of your churn rate requires more than just calculating the numbers.

Churn TLDR

› Churn is a measure of the value you deliver to that customer.

› It's usually more relevant after a year or two of being in-market with a product.

› Rule of thumb is that a 5 percent annual churn rate is acceptable for SaaS companies.

› If your churn rate is high early in your startup's lifecycle, you should consider a pivot.

› To really understand what's going on in your company, you should look at your churn rates in combination with changes you've made to your business.

CLV/CAC

Comparing Customer Lifetime Value (CLV) with Customer Acquisition Cost (CAC) helps assess the viability of a company's business model and its go-to-market strategies. The higher the CLV/CAC ratio, the greater the overall return on investment will be per customer.

CLV/CAC = total fees you receive from a customer during the "lifetime" of that customer ÷ cost of sales and marketing to acquire that customer

Calculating CLV only becomes more math than myth when your company has reached some level of scale with customers you've been selling to for a few years. How do you know what your CLV is? Articles will tell you to do this:

CLV = Average order value × Number of transactions × Average length of the customer relationship (in years)

For startups, getting these numbers is not realistic, because you don't have enough data yet to know the average length of your customer relationships in years. Instead, you can estimate CLV using your churn rate, which you're more likely to have earlier on in your startup's lifecycle.

> CLV = Average revenue per customer (monthly or annually) ÷ Churn rate (monthly or annually)

It's nice to watch your CLV to make sure it's growing over time. But you can't really benchmark CLV alone against other SaaS companies. This is where CLV/CAC comes in.

You'd think calculating the cost of acquiring new customers wouldn't be that difficult, but it's rife with interpretation. What expenses should go into "acquiring new customers"? All of marketing and sales? Do you include product marketing? Is customer marketing in your marketing budget? Does your sales team support renewals as well as new customers? Should you include all of the "overhead" included in supporting sales and marketing, such as office space, legal services, IT services? You'll have to decide.

Once you do reach some kind of détente on what to include in the cost of acquiring new customers, you just divide that by the number of new customers you acquired during the period you calculated that cost for.

> CAC = Total cost of sales and marketing during a period ÷ Number of new customers acquired during that period

CAC can vary heavily by industry, and even within an industry, it can vary significantly by business model. Many businesses have published CAC by industry statistics, including a great one published by First Page Sage called "Average Customer Acquisition Cost (CAC) By Industry: B2B Edition."

Also, how much you spend on acquiring customers tends to change based on the economic climate. Between 2018 and 2021, many companies could spend their way to secure customer growth. That led to a slew of high growth, unprofitable companies, driven by an over-exuberant investment climate. Sadly (and completely predictably), it didn't last.

OK. The hard math is over. Now, just divide your CLV by your CAC.

General wisdom suggests that a good CLV/CAC ratio is 3:1. And the higher it is, the better it is. If your business has a 5:1 ratio, you're building something great!

A low CLV/CAC ratio definitely means trouble — you aren't generating enough revenue to cover the amount you spend on acquiring new customers.

If your CLV/CAC is low, it could mean one of many things:

1. The most obvious conclusion is that you spend too much on acquiring new customers — your go-to-market motion is inefficient. This is where most CFOs go first.
2. However, it could mean that your product isn't providing enough value to your customers. Customers use it for a while and then move on — lifetime value is short.
3. Alternatively, it could mean that you didn't price your product appropriately for the value it does deliver during the period your customer needs it.
4. Or it could be that you're facing strong competition and your product isn't sticky enough. It's easy for your customers to switch from your product to your competitors.
5. It could be an indication your customer success and support functions are weak. Once the initial sale is done, you aren't continuing to support your customers' long-term success with your product.

Some confuse a good CLV/CAC with an efficiently run business. But obsessing about one ratio is never the full story.

CLV/CAC TLDR

› The higher the CLV/CAC ratio, the greater the overall return on investment per customer.
› Usually more relevant once your company has achieved some measure of scale.
› Rule of thumb is a CLV/CAC ratio of 3:1.
› A low CLV/CAC could be an indication of several different operational issues. Dig deeper.
› Don't confuse a high CLV/CAC with an efficiently run business.

Mistake 2: Building the Wrong Minimum Viable Team

As an entrepreneur reading this book, you likely understand that developing a new product or service involves multiple iterations, pivots, and extensive changes. And as you progress from concepts to prototypes to minimum viable working versions to products ready to scale, the team's required skills will inevitably also evolve. The initial stages of a startup require people with a certain set of core strengths, and it can be difficult to accept that the people you initially envisioned — maybe people you trust who you've known professionally in a different capacity — may not be the best suited for the job.

Archetypes and the Entrepreneur Archetype

To address this common hiring oversight, let's introduce the concept of archetypes. Archetypes are a representation (with a short, sweet name you can remember) of a set of characteristics that you use as a model to evaluate people. The beauty of archetypes is their flexibility; there's no standard template, and the traits you choose to look for are subjective and tailored to your project's needs. For any role you may be hiring, identify what archetype you're looking for and create your interview around that. For example, when Maria hires product managers, some of the archetypes she's used in the past include:

> Scalers: a passion for process, operations, and systems-thinking

> Growth-minded: highly competitive, data-driven, creative

> And of course, Entrepreneurs

In your launch (and pivot) phase, it's crucial to hire individuals fitting the Entrepreneur Archetype. It's not the same as simply hiring people with entrepreneurship experience, no matter how positive or negative the outcome. Here, we focus on the core strengths and traits that typify entrepreneurial individuals:

> **They're extremely comfortable with ambiguity.** To survive the chaos and extreme ambiguity of a startup in the Launch Wave, entrepreneurs are smart, intuitive, and able to function without knowing all the facts, solving problems creatively without frameworks or processes. One way you can recognize them during the interview process is that they won't ask a lot of questions about the scope of the role and responsibilities.

> **They're well-rounded.** You need people who can connect the dots at a time when you don't even know where the dots are going to be. Except for certain deep tech or specialized companies, you want to hire people who have a range of experiences, who can find opportunities within the opportunity, stay alert, and move fast. Hiring someone who has worked on the same types of products in the same domain for a long time can be a red flag, since they may be set in their pre-existing knowledge, and everything will look like a nail for their hammer.

> **They're proactive and risk takers.** You need people who take the initiative even when good outcomes aren't guaranteed. Startups have few playbooks, and most of the time you depend on people to just do things on their own without being told. Look for signs of initiative in their personal lives or how they approached their roles. Are they curious? How did they get into their different roles — was it part of the natural growth carried by inertia or were they their own agents of change? Did they do something unexpected?

> > They're hard workers and roll up their sleeves. You need people who can wear multiple hats and have a generous understanding of what their contribution means. In a startup, one day you're designing a feature and the next you're printing shipping labels to literally ship a workaround that unblocks a customer. No prima donnas.

> > They work with enthusiasm and grit. You're going to spend a lot of time together, under a lot of pressure. Things may go well, but they'll also go sour often, so you need people who display stamina, grit, and a smile while they go through the thick of the storm. Look for signs of committing to something hard and seeing it through. Has this person jumped from job to job every year for the past ten years? That's a red flag.

Darrell (UserTesting's co-founder) fit these traits perfectly. He had a background in finance but moved on to launch a pioneering website back in the days of Web 1.0, the first version of the internet. He was continuously curious about new trends and felt comfortable (as comfortable as an entrepreneur should get) launching another startup when he heard the clear pain point and unique solution his soon-to-be co-founder, Dave, described. Darrell and Dave did everything in the early days of UserTesting, from building and testing the product to managing the accounting and ordering office chairs. The two co-founders enjoyed working together, and decades later, they're still close friends.

Three Key Skills for Your Team

Since this is the hiring chapter, let's spend some time on the three key skills you'll need in your Minimum Viable Team: the product thinker, the builder, and the storyteller. We aren't saying that you need three people to create a startup, but you need those three skills across the founding team.

PRODUCT
THINKER   BUILDER   STORYTELLER

1: The Product Thinker

A good product thinker brings product and design sense with an ability to craft useful, simple, and business-aligned solutions. If you don't have these skills in your founding team, it'll become very apparent when the startup has built a technology marvel that fails to find customers. You need someone who deeply cares about the problem being solved, is empathetic and curious, and is able to put themselves in the shoes of your users or customers to identify problems worth solving. For a startup, the product thinker is like a relentless investigator, constantly talking to users to extract the insights and drive requirements necessary to build a solution. This research isn't something you can outsource. We have yet to see external research result in any insights that we didn't know before or that were rooted in any sort of reality of our situation.

This product thinking is usually called product management, which is a relatively new function. You couldn't study product management at universities until quite recently. Nowadays you can learn the basics everywhere, but please don't enlist an entry-level product manager for your startup. Building the skills to be a good product manager takes a broad range of experience across functions and often domains, plus years of experience. Did you know that some of the companies that made product management a staple of product development, like Meta, hire very few entry-level product managers? And when they do, they're required to complete three supervised rotations in different teams to intensively build that much-needed broad experience.

2: The Builder

Many startups are rooted in a technological innovation that requires domain expertise to build. To be successful in this phase, domain expertise isn't enough. As an entrepreneur friend of ours says, "You may not need to invent anything." What that means is you need someone who's able to tinker and can creatively use existing solutions to prove concepts and keep moving fast.

Because you're in the launch phase, you need an entrepreneurial archetype as your "engineering" lead — whether that person is engineering a technical solution, a service offering, or a new formula for protein bars. What's important is that your builder is less concerned with building something elegant or scalable and more concerned with proving hypotheses as fast as possible. Having a builder who has product-thinking skills helps tremendously because they can anticipate architecture decisions that will save so much time in the future. During the research phase, try to bring your engineers along when talking to customers — understanding the problem space and where the requirements are coming from will make your solution more likely to succeed.

We've seen startups try to outsource their engineering altogether. This is a bad idea. There are a number of legal reasons and protections that you should be mindful of, but even if all of that were documented, one huge reason to not outsource all of your development is the loss of speed of iterations.

> One huge reason to not outsource all of your development is the loss of speed of iterations.

In the beginning, you receive a lot of information from your customers, suppliers, and partners that helps you to understand what needs to be changed. Your builder needs to be able to use this information to adapt very rapidly. If you put someone in the middle who needs constant updates on context, it will crash your boat. To deliver on the next iteration of your product, you need clear and fast communi-cation and feedback loops between your customers, your product thinkers, and your builders.

3: The Storyteller

Half the job of creating a good product is surviving to see another day. And that's what the storyteller buys you when you don't have much to show yet. Your storyteller creates interest and gains support from customers and investors. They show not only your progress today, but how it will accrue and become something amazing in the future.

It's not sufficient to build a product and have a strategy to conquer the world — you need to tell the story. Without a storyteller, you may be okay at the beginning because visionary customers might buy into trying your product. But often these visionaries are just lurking in innovation departments that specialize in demos, trying new things fairly disconnected from the real needs of actual users. You'll need to put together the story of your product evolution, from the bare bones version you're pitching today to the future roadmap and benefits, so the real users will bet on you and start adopting it.

What to Care About When Hiring

Based on the hundreds of interviews from our entrepre-neurial backgrounds, here are the things you need to care more — and less — about when hiring.

Entrepreneur Archetype: *Absolutely Care*

Archetypes are rooted in core values and strengths that are hard to teach on the job. If you don't hire the right archetype in a role, there's a high chance everybody's going to be miserable along the way, and you'll end up needing to hire someone else down the road.

This is a mistake we've seen many startups make repeatedly. Instead of focusing on the archetype you need for the phase your startup is in, a connection (a friend, a smart ex-colleague, etc.) gets hired. Or even worse, someone gets hired for the title or clout that they could bring to your startup. For example, hiring a long-standing head of product from a big company to help you find product-market fit isn't always a good idea. The strengths that made that person rise in the ranks of the corporate world — organizational design while scaling a team or collaboration with other departments in complex multi-goal environments — are most likely not the ones you need right now.

Domain Expertise: *Care Less*

The majority of people we've hired for businesses in the middle of their Launch Wave have been "generalists." This means the person can do, and do well, a variety of job functions, but we don't expect them to be an expert across those job functions or know the ins and outs of the specific industry, technology, or product of the business. There will be cases in which you'll need very specialized knowledge, such as highly regulated environments or very complex technology, but in most cases, a strong generalist will do. Especially in the world of technology where things change quickly, focus on making sure your hire can cover the work that needs getting done and that they're curious and able to learn quickly.

Domain Passion: *Care Less*

Passion for the domain is a very important trait for entrepreneurship, but we prefer it when people have a passion for solving problems and achieving success. Passion for a particular subject is overrated — it adds a dose of grit and perseverance in the face of inevitable setbacks, but it's not a guarantee for success in the job or something we'd hire for. Having worked in Silicon Valley — the land of startups — for decades, we couldn't keep track of the number of people passionate about a particular domain or project who were later enthusiastic to work on the next big thing.

People have a lot of interests, and engineers and product thinkers can get passionate about so many interesting problems. So no, having domain-specific passion isn't a prerequisite.

> Passion for the domain is a very important trait for entrepreneurship, but we prefer it when people have a passion for solving problems and achieving success.

Years of Experience: *Be Careful*

You'll need different levels of seniority for different roles. However, not all years in candidate resumes are equal — especially when they're coming from big companies, where exposure to complex problems with high levels of ambiguity isn't a given — nor is having the agency to solve them. You can't easily tell from a resume if a candidate is able to exercise the levels of autonomy that a startup needs, so you probe for this trait during the interview process. Ask about a project retrospective they led and how decisions were made. What was imposed and what was the candidate's role in the analysis and decisions? Watch out for answers to interview questions that begin with, "We decided to . . ." Ask, "Who is 'we' in this case?" Use the interview to help you calibrate the candidate's years of experience with the level of autonomy you'd expect.

The Right People for the Phase You're In

One of the mistakes entrepreneurs make is hiring someone for the needs of today (Launch) and tomorrow (Scale), because that's rarely the same person. You'll have different needs along your journey, but you need to hire (and fire) for the phase you're in. This implies that in the future you may need to replace people as your needs change. Some people are avid learners and flexible, so they'll be able to adapt and grow with the company. Others will have to go to make room for the newcomers with the skills required at that time — and that may include yourself.

> You'll have different needs along your journey, but you need to hire (and fire) for the phase you're in.

In startup culture, grit and persistence — and the accumulation of vesting stock at the potential gigantic exit — might cause people to extend their welcome. Other people lack the self-awareness to realize it's not working. Not recognizing the mismatch of skills will make everybody unhappy, and cutting ties in this situation is the healthiest thing to do for everyone involved.

Often when you must let someone go due to performance, it's not about their performance in general — it's about their performance in that role at that moment. Usually, there's a mismatch of archetype with what the role requires. Developing those core strengths is hard, even in normal conditions. It's even harder in a startup (a pressure cooker disguised as a company), where there's an almost nonexistent training budget and a lack of coaching programs. That person would be much happier in a different role in which their core strengths match what the role requires. A startup can teach you a lot very fast about an industry or how to perform a particular function, but it's not the place to develop new core strengths. What you hire is what you get.

Key Takeaways

---

1   Hire for the right archetype rather than for seniority, domain expertise, or domain passion. You're looking for entrepreneurial types who are comfortable with ambiguity, proactive, and well-rounded.

---

2   Your team should include people who can fill the role of product thinker, builder, and storyteller.

---

3   It's important to hire for the phase you are in, not the phase you hope to be in later. The skills required at each phase of your development are different, which may require you to replace members of the team as your needs change.

Mistake 3: Being in Love with the Solution

Becoming too committed to a solution too early is one of the most common mistakes entrepreneurs make as they navigate the Launch Wave. It's an indication of an unwillingness to rapidly evolve their understanding of both the problem and the potential solution.

There are two primary reasons for this mistake, which may occur independently or in combination. The first is *ego*. Ego is not your friend in the world of business, and in the world of startups it may be deadly for your company. In a startup, ego shows up when the founding team becomes emotionally attached to a solution. If the solution is particularly clever, innovative, or just hard to build, the founding team might overvalue the technology's potential impact on users and struggle to reconcile this with reality.

> Ego is not your friend in the world of business, and in the world of startups it may be deadly for your company.

The second reason could be that the team hasn't conducted sufficient user and market research, or having done so, ignores the findings due to confirmation bias. Confirmation bias is a trick the brain plays on us, seeking or selecting information that confirms our preexisting beliefs or hypotheses. That's also the ego not wanting to listen to the reality of what customers are telling you.

Here's an example of how these two factors can work against you. We worked with a hot Silicon Valley startup founded by highly talented scientists transitioning from academia to the starkly different world of business. They had developed a novel biotech procedure with theoretically numerous applications. Initially, they focused on one application that seemed innovative and newsworthy — a seemingly sensible solution to a problem. The positive press and venture capital interest seemed to also confirm that this capable team was onto something. However, once they started to engage with customers, the situation changed. Despite initial curiosity from many Fortune 500 companies that led to meetings and pilot projects, deals were not materializing, and interest faded quickly.

Within the first few months, two major issues with the product became clear. The first was that the problem solved by the technology wasn't significant enough to motivate users to adopt a new tool; the customers' main concerns were selling more products and not being sued. The product provided value in an area that wasn't selling more products or preventing being sued. Hence, it was merely nice to have. The second issue was the proposed product was too expensive, operationally burdensome, and difficult to integrate into established company workflows. The combination of these factors clearly pointed to a need for a pivot in either the problem to tackle, the solution, or both.

Typically, this is the point where founders pivot. But in this case, it wasn't that straightforward. It took months for the founders to accept this reality. The initial press, the attention and compliments for the cleverness of the solution were dragging the decision. This team was in love with the solution they had originally adopted. They only came to terms with the situation when the inescapable realities of business plans and business projections started to hit home. Once this occurred, they did begin to pivot, but it was months from when the problem was first identified. This delay could have been fatal for the startup. Fortunately, they still had some time and funds in the bank account left to ideate and relaunch into more fitting waters. The stars aligned, and they're still in business.

Why We Fall in Love with Our Ideas

It's the most natural thing for people to fall in love with their own ideas and solutions. We justify our attachment to them in a number of ways.

It's Technically (or Design-wise) Superior

Engineers and designers take great pride in building and designing clever solutions. Everybody wants to do a great job, perfecting their craft as product thinkers and builders. Solving complicated technical challenges or designing simple and delightful user interfaces is fantastic when you're solving the right problem, or when all the other parameters that make a product work are there. But if you haven't found your product-market fit yet (which is most likely the case in the early days of your startup) being attached to a particular solution because it's superior will be a drag for a much-needed iteration or pivot.

It's Innovative

Similar to the pride of building something superior is the pride of building something new. Innovation doesn't win contracts or adoption on its own. Just as people don't necessarily buy a product because it's superior, they don't buy a product because it's novel or innovative. And don't confuse your early fans, the innovators, with proof that you have a solution that will support a successful business. Innovators are people who will try nearly anything that's new. But the novelty effect is not long-lasting, and what innovators do isn't transferable to bigger cohorts, like the early adopters or the early majority. You can find more information about the different cohort types in the Sidebar at the end of **Mistake 7: Too Attached to Pivot.**

> Don't confuse your early fans, the innovators, with proof that you have a solution that will support a successful business

---

People Love Talking About It

You might not have paying customers yet, but hey, they do love talking to you, especially the innovation team. You attend panels, fireside chats, and present in many conferences. Also, you're in the press. You caught a trend or a particularly interesting angle for reporters, and they printed the stories you pitched. It's such a badge of honor to be in the news, or to tell your relatives that you have meetings with Fortune 100 companies who are interested in your product. But you didn't start this business to collect accolades. And unless all the other requisites to build a great product are met, this attention will only do one thing — stroke your ego and make it harder to move from falling in love with your solution to falling in love with the customer's problem.

> Unless all the other requisites to build a great product are met, customer and press attention will only do one thing — stroke your ego.

It Was So Much Work to Build (Sunk Cost)

Economic theory says that sunk costs — which are investments that have already been made — shouldn't be relevant to future decision-making since they are non-recoverable. In practice, it's psychologically challenging to let go of an investment, driving people to make poor future decisions based on bad prior investments.

Whether you've built the right solution or the wrong one, basing future decisions on past sunk costs is only going to increase your chances of sinking further. And if you were too in love with your solution, then the sooner you realize it, the sooner you can minimize further losses. The longer you wait, the more you invest in your existing solution, the more you fall in love with it, and the harder it gets to change direction if you need to.

If you realize your love for your solution is preventing you from cresting your Launch Wave soon enough, no harm done. The problem comes when you resist seeing it and double down on making it work. How can you avoid making this mistake? That's what we'll cover next.

How to Tell if You're Just Being Stubborn

There's a fine line between grit and stubbornness. How can you tell the difference? Here are some practical ways.

List All of Your Assumptions

As you start moving towards ideation to solve a particular problem, you'll be making tons of assumptions. The brain works in mysterious ways, and we're all subject to a number of biases. Confirmation bias is particularly dangerous when you're trying to build a new product or service.

Through the months or years that you've been working on your project, you've probably made a lot of assumptions. Some of them might have already become pseudo-facts in your brain due to confirmation bias. Combat this by very intentionally listing all the assumptions you're making. A very simple yet powerful technique to use is the "We believe . . ." statement. It goes something like: "We believe that if we do X, customers will do Y, because of Z." Here's a "We believe" statement that could've worked for Airbnb: "We believe that if we provide a service that offers available rooms in people's apartments for rent, travelers in town for a conference will pay to stay in that room, because it offers an affordable lodging option when hotels are mostly full."

By listing all your hypotheses with a statement that starts with, "We believe," you reinforce the message that it's not fact; it's an assumption you need to validate, as quickly and as cheaply as possible. Giving more details on the customers' actions and the reasons you think they'll take those actions will help you dig deeper into understanding your users' motivations and needs. Make sure each of your hypotheses is validated. If you're substituting facts for assumptions, you'll be drinking the proverbial Kool-Aid.

> By listing all your hypotheses with a statement that starts with, "We believe," you reinforce the message that it's not fact; it's an assumption that you need to validate, as quickly and as cheaply as possible.

Focus on the Customer and the Problem

Instead of continuing to obsess over the solution and adding more features, focus on validating your assumption of your customer's problem. Go back to the basics of understanding your customers and prospects, starting with the industry. What's the industry value chain? What are the major forces that shape the market? How do they make money? What does your customer's CEO worry about (and put budget against)? How are customers currently solving these problems? Are they happy about it?

You should spend a significant number of hours just understanding the landscape of the industry that you're entering. If you're already an expert, you're a step ahead. If you aren't, you'll need to drink from the fire hose: attend conferences, read industry news, research reports, letters to shareholders, and much more just to get a baseline of your customer landscape. Most importantly, meet with and listen to your prospects and customers.

Listen to Customers

Note that we're not saying talk to customers — listen to them. And the reason we emphasize this is because you're in love with *your* solution. We bet that after quick intros and pleasantries in customer meetings, you go into selling mode. Instead, what you need to do is ask a lot of questions. The first order of business is to make sure you've identified the right problem to solve, and you're validating your hypothesis. Be careful your confirmation bias doesn't creep into your questions. As they say in courtroom dramas, don't lead the witness. Instead of asking, "How do you like our newly launched advanced AI feature?" ask, "Which of our new features do you like and why?" A great book on how to find out exactly what customers care about — and what to say and not to say in those conversations — is *The Mom Test: How to Talk to Customers and Learn If Your Business Is a Good Idea When Everyone Is Lying to You* by Rob Fitzpatrick.

Don't Listen to the Wrong People

Some people you talk to could be quick to encourage you to go for it and keep pouring money into an idea you have. If these people aren't users themselves, you're likely being misguided by hype. There are three sources of feedback that may guide you in the wrong direction:

1. Your support system: In general, people don't want to let other people down, especially those they care about, and most of the time they'll only give positive feedback. Don't let encouragement from friends, family, or well-intended colleagues detract from your research with actual users. Everyone needs a support system in their lives, but your strategic decisions should be based on users' likelihood of using your product — not your partner or best friend telling you how smart your idea is.

2. The wrong type of customer: Another mistake in this area is listening to customer feedback from people who aren't decision makers, users, or buyers. Sometimes central teams such as technology selection or innovation teams will evaluate new products, but if they aren't a decision maker, user, or buyer, then their feedback can steer you in the wrong direction. Respectfully ask to talk directly with the users and, in the best-case scenario, get them to try your product in a real-world situation rather than a sandbox.

3. A single type of customer: Aim to diversify the types of people using your product. Different perspectives can give you insights and reveal pain points and needs that aren't shared but are extremely valuable.

Remember, the longer you double down on something you've already built that isn't working, the higher the chances you'll crash against the Launch Wave, probably by running out of money before having time to build the real deal. Breaking up with your solution is hard to do. It takes courage and a dose of humility, but the sooner you make the pivot, the better for everybody involved.

Key Takeaways

1 Don't let anyone other than real customers' feedback get in the way of what a good solution is or is not. Listening to friends and family, press, and non-users may stroke your ego, but it doesn't pay the bills.

2 There's a fine line between grit and stubbornness, and your ego may get in the way of recognizing when it's time to evolve your solution.

3 When in doubt, go back to the basics, listing all your assumptions and deeply listening to customers. Be aware of your confirmation bias both interpreting their answers and in the questions you ask them.

4 Breaking up with your solution is hard to do. But doubling down on what isn't working is just adding to your sunk costs.

Mistake 4: Building the Wrong Minimum Viable Product

After you've done your initial research, you should have a clearer understanding of the market and industry you're entering. You should be familiar with your target audience, as well as their current challenges and potential future obstacles. You should also know the way your potential customers operate, the tools and workarounds they use, and their pain points. After this thorough research, let's assume you've identified a problem important and big enough that demands a solution. You believe you can build that solution and create a business around it.

The next step is to integrate this knowledge into designing the right solution, building and testing the many assumptions and hypotheses you've made along the way, starting with building a minimum viable product (MVP).

Your Product in the Real World

User response to a certain product, when presented as an abstract idea or visual representation, is different from the behaviors shown when interacting with a product in reality. That's why researchers deeply mistrust self-reported survey data. People's words can't be trusted. It's not that people lie (well, sometimes they do), but rather it has to do with unconscious motives, conflicting views of the self, lack of constraints when evaluating future behavior, and so much more. You can only really know how your users will react to your product when you put it in front of them, and they use it in the context it's intended to be used in. If you want to dive deeper into this aspect of user research, then a great book to read is *Competing Against Luck: The Story of Innovation and Customer Choice* by Clayton Christensen. It introduces the "Jobs to be Done" framework, which is used to understand the underlying reasons customers buy or use products, focusing on the "job" they're "hiring" the product to accomplish in their lives.

> You can only really know how your users will react to your product when you put it in front of them, and they use it in the context it's intended to be used in.

In *Competing Against Luck*, the case study of how McDonald's increased their milkshake sales remains one of our favorite examples. Initially, McDonald's tried varying flavors and textures, but sales remained flat. It was only when they understood the real reason people bought milkshakes — to stay entertained during commutes — that they found success. People needed one-handed snacks that lasted long enough to cover their commute and didn't create a mess in the car. With these insights, they made adjustments that solved for those specific problems — such as making the milkshakes thicker (so they lasted longer) and adjusting the sales experience (to make it easier and quicker to purchase a milkshake) — and sales finally rose. Voila!

Genuine feedback, market reception, adoption, and hypothesis validation occur when users interact with your product, for real. And that's what your MVP is for. It's a vehicle to test hypotheses quickly and with minimal effort, enabling rapid iteration on the learnings to meet actual customer needs and create a viable, marketable product.

What's Wrong with Most MVPs?

Search "minimum viable product" on the internet, and you'll get over a billion search results. What can we say that hasn't already been said? And yet, despite all the wise words out there, most startups and new product development teams get their first product attempt wrong. This happens because it's extremely hard to strike the right balance in defining the MVP features appropriately — not so many that they make development and iteration slow and the product too complex for the stage you're in, and not so few that you can't provide a viable solution to test your hypotheses because your MVP lacks sufficient functionality.

Overloading an MVP with features slows development and potentially increases the sunk costs of something that still needs significant iteration. We usually see this when products try to cater to various user types with different needs. This results in a feature-heavy product that loses usability. Your MVP will be complex and unclear, and it'll take much longer to build. We also see issues with overengineering MVPs. When you expect your product to knock it out of the park, you may be tempted to scale it too soon, building for high traffic, massive storage, and the reliability necessary for millions of people to use it at once. But remember that nine times out of ten, you're not building the right thing this early, so overengineering your MVP will not only slow you down, but it will also create more sunk costs when you inevitably need to make changes as you iterate.

Conversely, not including enough features, or the right features, can be just as detrimental. If you overemphasize the minimum part of your MVP, you might miss delivering essential features that reduce your product to something less than a viable solution. Or you could cut corners in the user experience, making the use of your MVP frustrating for your users. When you don't scope your MVP appropriately, you won't know whether the negative feedback you're getting is because you missed solving for your user's real pain point, or you failed to include critical features, or you included to many features that confused your user, or your user just had a poor user experience — leaving you without the learnings your MVP was built to deliver.

To strike that balance between features and viability, focus on the core functionality that addresses the primary user needs with a positive user experience, without overcomplicating or oversimplifying the product. Easy, right? So how do we do that?

Your MVP is a vehicle to test hypotheses quickly and with minimal effort, enabling rapid iterations.

Focus on *Some* Customers

Even if your startup's strategic goal is to conquer the world (which it absolutely should be) and address a multitude of problems for various user groups, when it comes to your MVP, narrow your focus to a specific type of customer. Introducing new products to the market and convincing people to adopt them is challenging enough. This challenge multiplies when you attempt to cater to the needs of multiple user segments simultaneously.

Instead, concentrate your efforts on one specific customer type, develop superior solutions for them, and minimize the number of variables when gathering insights. You can always expand your audience or pivot to a different customer type after gaining valuable insights and traction. Your MVP shouldn't attempt to cater to everyone's needs right from the start. In fact, attempting to do so significantly reduces your chances of success. You're more likely to try to add too many features or overengineer, prolonging your development time and increasing the complexity, dependencies, and overhead. It also limits your opportunities for iterations while increasing sunk costs when adjustments become necessary.

Deliver the Core Value

An MVP isn't merely "building something." It must be viable. You should be able to validate whether the core value proposition your product offers resonates with users. In addition, given that the MVP serves as your primary learning tool, you should be able to validate the core hypothesis of your product plan. If your strategy depends on users taking a certain action, then ensure this action is included in your MVP and that you've built a way to validate when users perform the action.

We often see situations in which an MVP is deployed but doesn't get traction, yet the team remains puzzled about the reasons. It could be because the core value proposition doesn't resonate with users, or because a critical enablement feature wasn't built, and it created too much friction to adopt.

To prevent this problem, it's wise to go back to your original list of product hypotheses (the list of statements that begin with, "We believe . . .") and compare it with your MVP feature list, ensuring your key hypothesis can be validated with the features you're building into your MVP. Another technique is to rank the features you want to include. Every single feature needs to earn its way in either by being part of the core use case or by helping validate a key hypothesis for your business. If neither of these are true, then the feature is a "nice to have," and you should therefore (painfully) postpone it.

> Every single feature of your MVP needs to earn its way in either by being part of the core use case or by helping validate a key hypothesis for your business.

There are numerous other areas in which you can be resourceful and prioritize development speed over development cycles. For non-differentiated parts of your solution, get off-the-shelf modules. There are probably designers that have great ideas for a new way to implement a timer, but unless your startup is building an intelligent, quantum-computed, next-generation timer, it's not worth the time and effort.

While this might not be intuitive, you should also do things that don't scale — such as relying on manual work and operations to fulfill some activities that could be automated but shouldn't be until you clear your Launch Wave and are ready to scale. And your engineering team should resist the temptation to overengineer your infrastructure to handle massive usage. It's highly likely you won't require such scale immediately. By the time you do, you'll have accumulated a wealth of insights about your products and user behavior, necessitating a re-architecting of your infrastructure anyway.

Don't Deliver Too Fast

Yes, we just told you to focus, but there are areas you can splurge on because they're worth it: design/user experience and metrics.

Great design and user experience of the core value proposition are worth it. If your MVP is low-quality, breaks, or is hard to understand, your users may have a bad reaction and not engage. Remember that first impressions are lasting ones, and they're usually made rapidly when first experiencing a product. If user experience was bad, then you wouldn't know if your users lost interest because of the experience or because of the core value proposition. You may attempt to re-engage them later by fixing things, but you can only have one first impression. Make sure your MVP has gone through the right level of testing and usability analysis before you launch it.

If you don't have a plan to gather feedback and learnings, your MVP will be useless. Feedback can be both quantitative (making sure you've invested in observability and the ability to look at your metrics) or qualitative (with surveys or interviews lined up to gather customers' feedback). This is one of the most precious values of your MVP.

Past the First MVP

We emphasize identifying as quickly as possible when your MVP isn't working, and you consequently need to iterate on it. What we aren't emphasizing on during your Launch Wave is creating a scalable, full-featured product. Remember, your MVP is a vehicle to test hypotheses quickly and with minimal effort, to quickly converge on a viable, marketable product. Continue iterating with urgency until you're sure your MVP has connected with a possible product-market fit. If you confirm that you can sell it, then you're heading to the Scale Wave. Otherwise, you're back to iterating on your MVP, or perhaps having to face the Pivot Wave.

MVPs Done Right

The path UserTesting had to take to overcome its Launch Wave, and eventually also conquer its Scale and Exit Waves, wasn't as easy a journey as perhaps our story might indicate. The founders launched around the time of the 2008 financial crisis that resulted in banks that were "too big to fail" failing, stock markets tanking, and investors taking a break from investing to lick their wounds. Dave and Darrell weren't able to raise any material amount of money for a few years. Darrell recalls, "The only thing that got us through this period was an incredible amount of persistence, the fact that we really enjoyed working together, and that our customers continued to tell us they were amazed by the quality of feedback they were getting and how quickly and easily they were getting it."

You'd think the co-founders had built a robust, full-featured product to hit on that kind of customer love. In fact, they had a vision and strategy for what they could build — one day. But during their Launch phase, instead of listening to the wrong people — in this case, the UX thought leaders who would've had them thoroughly vet people before allowing them to sign up and give feedback — their MVP featured a simple online form with the minimum amount of friction necessary to vet participants. That did the trick.

Key Takeaways

1. The right MVP is a vehicle to test your product in the real world, enabling you to act quickly on your learnings to meet actual customer needs and create a viable, marketable product.

2. You're looking to strike a balance in your MVP between building enough to learn from and moving quickly. Too many features can take too long before validating the core idea, but an incomplete MVP will fail to solve the problem or provide conclusive feedback.

3. Continue iterating with urgency until you're sure your MVP has connected with a possible product-market fit. If you confirm that you can sell it, then you're heading to the Scale Wave. Otherwise, you're back to iterating on your MVP, or perhaps having to face the Pivot Wave.

Mistake 5: Scaling Too Soon

The last mistake of the Launch Wave is scaling sales and marketing too soon. Finding product-market fit includes more than creating a product people want to buy. It also requires you to figure out how to build a repeatable sales and marketing motion.

If your team has identified the right problem to solve and has built an MVP, yet you still aren't seeing solid traction, then it's tempting to believe your problem is that you need more of a pipeline of people to sell it to. That would drive you to believe your next step should be to hire a salesperson or two. But most of the time, your problem isn't a pipeline problem — rather, you still haven't found your product-market fit. Perhaps your MVP is the problem? Perhaps you're focusing on the wrong customer segment? Perhaps you haven't learned how to position your product for the customer segment you're selling to? New sales hires aren't going to find out what you need to do differently. That's still your job.

> If in this wave you aren't selling, then new sales hires aren't going to find out what you need to do differently. That's still your job.

Don't Build a Sales Team Yet

You might think a sales team will help you better reach prospects, but really it can stand between you and your customers. Here are some of the tactical things that happen when you build a sales team too soon.

They'll Filter the Feedback You Need to Hear

If your salesperson is front-lining with your potential customers, then you won't be the one getting firsthand feedback from customers. In the early days of a startup, firsthand feedback and the ability to ask follow-up questions in the moment while talking to your prospective customer are invaluable. As an entrepreneur, your aim during this wave should still be focused on learning — learning how to create a repeatable sales motion in this case. When you hire a salesperson, their focus is on revenue generation. And while they might close a deal or two, it might not be the deal that drives profit, or the deal that renews, or the deal that's repeatable across a segment of customers. And if they can't close a deal, then they're less likely to take the time to understand why. Because they're incentivized to go to the next prospect rather than spend time with the one that isn't going to buy from you.

As a founder, you're the only one who will take the time to talk to the prospect that isn't easily convinced to buy. Your job is to find out why your product isn't hitting their pain points and what their adoption challenges are. Your customers aren't going to come out and tell you — you'll need to probe subtly. Founders are the ones who have the most context on the industry, the product, and the value proposition that's being built; they're also driving the iterations and roadmap of the product. For them, these first-hand conversations with prospects are crucial. When that feedback goes through a salesperson, most likely it would be shallow, filtered, and sometimes biased.

They'll Hide the Shortcomings of Your Product

It's a little cliché, but a good salesperson — with their charisma, enthusiasm, and persuasive arguments — would be able to sell almost anything, especially a pilot project, to your prospects. If you hire a salesperson who's very good at their job, then their ability to sell can obscure the real shortcomings of your product. While it's thrilling to see signs of customer interest, it could be bad news if their onboarding isn't supported by the realities of the product and its roadmap.

They'll Push You to Lose Your Focus

Because a salesperson's instinct is to sell, they'll try to make a sale, any sale, even if it's not what your company or product stands for. This usually manifests in requests for customized features that Customer A is willing to pay you to develop. If the custom feature is unusable for any other customer, then building it will just distract your precious startup resource from building the right MVP. Making a sale and having revenue are tempting, and your flashy new salesperson will be a very strong and influential advocate inside your company pushing you to invest in this customization.

Sometimes, if customers are willing to pay for the work, startups will take it on, so they can use the money to bootstrap their business without having to raise money from external sources. However, taking on services work if you intend to build a product company should be a very intentional decision. You either do sideline contract work to bootstrap, or you remain focused on the endgame. You should be making these decisions proactively and very intentionally, not letting a salesperson make them in a reactive way.

They'll Overpromise and Create Too Much Hype

There's yet another reason to avoid putting salespeople in front of your prospects too soon. In their effort to sell, an enthusiastic salesperson will often lean to overselling — pitching your MVP as it could be when it becomes a full-fledged product with all the bells and whistles to customers. Once you've found your product-market fit, and you're ready to ride your Scale Wave and iterate fast on product development, this can be a good thing. But overdoing it during your MVP despite the reality of the state of your current product will leave you with early customers who feel you've overpromised and underdelivered. This leaves you in catch-up mode, failing at expectations and creating negative sentiment.

Don't Do Partnerships Yet

Partnerships are another way you may be tempted to scale your reach to potential customers. In this model, you lean into another company, with products usually complementary to yours, to sell your product. These other companies are usually bigger and more established than you are. While it can be exciting to get interest in your tiny startup, you may be lured into thinking all your distribution problems can be solved since you're providing functionality they lack. Well, think again. For that partner, you're probably just a checkmark on one of many checklists.

The other major drawback is that they're big and have plenty of resources to do these deals: negotiate, ask for meetings, tons of legal back-and-forth, etc. You have neither the time nor the money to invest at that level. The return on investment (ROI) will most likely be terrible for you. We'll get into more detail about how to avoid partnership mistakes (and how to do partnerships right) in **Mistake 14: Partnering as a Side Hustle.**

Don't Scale Marketing Yet

There's still more you should avoid doing as you face down your Launch Wave. You should not scale your marketing at this point either. Some thoughtful marketing materials can position a startup as a little bigger and more established than it really is, adding to its credibility. But creating beautiful and expensive websites with menus and submenus, creating a repertoire of half-baked case studies, and publishing press releases that lack substance are just resource drains for startups in their Launch Wave. You will end up rebuilding your marketing collateral and website several times in the first few years. You still don't know what marketing works best for attracting the customers you should be targeting.

What to Do Instead — Wait

Yes, wait. Because a startup's first customers need to be won by the founders, not by hired salespeople, partners, or marketing websites. These prospects and customers drive the most critical learnings, the deepest insights, the most valuable information to help you navigate this wave — but only if customer feedback is directly collected by the founders. Don't hire sales or marketing when you're still validating hypotheses, because the only way to validate them is by staying as close as possible to those first customers. It's you who should be out in the field at this point.

 First prospects and customers lead to so much learning and so many insights, that founders need to be directly involved.

There are a lot of resources available to learn how to best collect that feedback from customers, but for us, the most important thing you should pay close attention to is how your customers use (or don't use) your products. Look not at what they say, but what they do. The first level of feedback is usually overly positive, and when they show interest, it's usually quite superficial. This feedback, delivered in the abstract and without constraints, is often useless. Rich feedback comes when customers use the product's core value proposition in their day-to-day, and that usually involves an MVP, as we discussed in **Mistake 4: Building the Wrong Minimum Viable Product.**

Here's an example: If you compare the healthy living goals you made on January 1 in your shiny New Year's resolution with what you're actually doing by March, there's probably a stark difference. What we think we would do, or aspire to do, is not the same as what we do in practice once we face the constraints of the real world. For customers, it's the same. It's not that they're lying to you, it's just that their feedback can't be trusted in this context.

Then, Hire the Right Archetype

The biggest mistake on top of scaling sales too soon is attempting to do it with a "professional career" official, someone who's been climbing the ladder to Big Company SVP or similar and hasn't been doing "the work" directly for a long time.

We don't want to kiss and tell, but as with many of the examples in this book, this happened in one of the companies we worked for. The first salesperson hired was a sales executive from a Fortune 100 company. This person not only brought the clout of the brand name, but also knowledge of the processes of what needed to be done and beautiful slides to explain it all to the executive team. The team was dazzled. There were playbooks to be built, sales motions to activate, and so much more. There was only one problem. The company needed a salesperson, not a sales executive that expected to build a sales team. And this hire was the latter — expecting to train and lead others on how to take care of each part of the process they were building out. But the company was still working on getting over its Launch Wave. There was no one else to do "the work" this sales executive was creating. It took a year of an expensive salary and considerable effort without results before the company and the sales executive agreed to part ways.

Instead of hiring a "professional career" corporate sales-person, when the time does come to hire your first salesperson, look for an entrepreneurial archetype, as we discussed in **Mistake 2: Building the Wrong Minimum Viable Team** — someone who loves the idea of being out in the field, testing different sales pitches to see what lands with which customer segments, and thrives on the thrill of the hunt. Don't bend to the temptation of hiring a big exec because of their contacts. They could get you an initial introduction, but you still need a tenacious salesperson to learn, iterate, and drive the deal to closure.

Key Takeaways

1	Scaling your sales efforts prematurely can rob you of the valuable face-to-face feedback you can gain by talking directly to customers.
2	Partnerships and marketing efforts are also likely to backfire — or at least drain valuable time and resources — if you haven't first nailed your product-market fit.
3	When you're ready to add your first salesperson, make sure they are of an entrepreneurial archetype.

Launch Wave Wrap Up

The Launch Wave is an exhilarating moment for any startup. It's full of hopes, dreams, ups and downs, new shiny things, and learnings. However, crucial decisions will be made during this time that will make the difference between you clearing this wave or becoming part of the immense graveyard of startups that never sailed to the Scale Wave.

We want to make sure you start with the right set of ambitions for a successful business — future proof, big, and profitable — and that you hire the team that will allow you to weather the elements — with the right archetypes and at the right time. As you start developing your product, you'll also need to be vigilant about your own biases and your ego and let real customers' feedback guide you. And as you build the first prototypes and MVPs, strike the right balance on features, so you learn as fast as possible.

But between your Launch Wave and your Scale Wave, there's a challenge you'll more likely than not need to face. Let's talk about the Pivot Wave.

The Second Wave

Navigating a Pivot

You've launched! You have an MVP and customers and some revenue, and enough people like your product that you think you can keep on making it. However, if you're honest with yourself, you may realize you're sailing against the current and exerting a lot of energy to move forward with little to no progress. A 2022 poll of nearly 500 startups found that 40 percent of the founders who responded had previously pivoted their startups in some fashion to avoid failure.[1] A Swiss Federal Institute of Technology Lausanne university study found that as many as 73 percent of startups had to pivot.[2] Based on our experience, the higher number tracks.

If you're aimed in the right direction, then it'll feel like a strong wind is pushing you forward as you sail past milestones that previously seemed out of reach. If, on the other hand, you're about to hit the Pivot Wave, then it'll feel like you keep needing to push harder for "just one more quarter" to turn things around. The traction you need is always right around the corner. How do you know if that's true? And how long can you afford to wait to find out? If you aren't sure about whether you need to pivot your business, here are some indicators now might be the time:

> You've launched your solution and have less than nine months of cash in the bank, and you aren't seeing the traction you need to generate revenue or justify more funding.

> Customer feedback indicates that you have a nice-to-have product, not an essential-to-have product, which is reflected in your retention/renewal rates.

> You have a product your customers love, but you're realizing there isn't enough revenue among those customers to support the growth you want to see.

---

[1] "Why startups," *Skynova* (blog).
[2] Sanctuary, "Where to play," EPFL.

> Your product is highly technical, and you've been in-market for a year or even longer. All your "customers" are running pilot projects out of their innovation groups, with no sign of adoption by their core business.

> You have a "freemium" product and lots of users, but few are converting to paid users.

> A new technology or process launches that rapidly reduces or eliminates the niche your company serves.

> You keep setting milestones and never seem to meet them.

A pivot is not a straight line, and sometimes it takes many turns before you're able to position your company in just the right spot. In this section, Navigating a Pivot, we'll guide through some common mistakes that happen as you pivot to get to the Scale Wave.

A pivot is not a straight line, and sometimes it takes many turns before you're able to position your company in just the right spot.

The first mistake we'll cover confuses effort for progress. In **Mistake 6: Flailing in the Water**, we explore how many founders mis-takenly add new products and features in hopes of jump-starting a flailing enterprise, instead of determining a clearly focused plan to pivot that allows them to sail confidently in a new direction.

In **Mistake 7: Too Attached to Pivot**, we consider what happens when you're too attached to the stories you've been telling yourself thus far. By detaching from the assumptions you've been dragging along with you, you can ask yourself new questions that will allow you to spot openings you haven't seen before.

In **Mistake 8: Ignoring False Commitment**, we explore how to determine if you've really enlisted a sense of commitment to the pivot, or if your team is saying they're with you when in fact they're not. Many leaders focus on the execution of the pivot itself, without managing the change within their own organization. We'll show you how to overcome people's natural resistance to change, so they can fully embrace the transformative idea you've set forth.

And finally, **Mistake 9: Not Going All-In** demonstrates how you can sink your pivot by failing to go "all-in." A pivot is a change in nearly every part of your business, and successful pivots are fully committed company pivots. We'll show you how to make sure your entire enterprise comes along for the ride.

Story:
Hitching a Ride on a Different Wave

Hitch was a tech startup that spun off from a bigger company in 2020. It initially focused on "talent mobility" — software that helped fill roles or staff projects and gigs internally, instead of hiring from the outside, so companies could cost effectively retain and grow talent. Hitch found their challenge — most companies already had a learning and applicant tracking system (ATS), many of which were adding talent mobility functionality to their solutions. And even though Hitch's talent mobility solution was superior, customers appeared to be satisfied with the "good enough" features their competitors were offering for much cheaper — or even for free.

Heather had been an investor and board member of Hitch. When she was asked to join Hitch's executive team, first as the president and then as CEO in 2021, she was faced with a challenge. Hitch wasn't getting enough traction in the market segment they were selling to. Their product, though better, wasn't different enough to overcome the friction customers faced when considering buying yet another solution instead of just working with the features of a tool they already had. The good news was that Hitch's existing large-enterprise customers loved the product. Heather knew the company had something valuable, but Hitch wasn't succeeding in turning that value into growth. They needed more information, as well as an openness to change.

With her team, Heather started a multi-month listening and learning tour, deeply analyzing competition and how customers used their product. They engaged customer and industry participants with many open-ended questions.

Questions like:

> Where were they on the skills maturity curve?
> Were they moving or planning to move to a skills-based talent model?
> How were they assessing the skills of their workforce?
> How were they identifying skill gaps?
> How did they plan to upskill, reskill, or redeploy their talent?

When they asked questions this way, customers started nodding furiously, telling them how frustrated they were with their existing solutions, managing all the disparate skills data they had across multiple systems, spreadsheets, and in some cases, in a manager's head. They didn't have a way to harness the silos of information they had to grow and develop their workforce effectively. This seemed like a bigger problem that wasn't solved within the industry.

Having identified a bigger problem Hitch could possibly address, it was time to take inventory of Hitch's internal capabilities. Hitch had a small but strong data science team that had been working on building proprietary algorithms using a dynamic dataset of billions of global data points to connect skills, roles, and people. These algorithms could take all the disparate skills data across multiple systems in a company and clean, normalize, and deduplicate it into one unified skills ontology, and use that intelligence across an entire employee population to find and match people to career paths, learning courses, mentors, projects, and more.

Doing this inventory, Heather realized their functionality was more advanced than what the market considered a talent mobility solution to be. Hitch had built an open, dynamic, AI-driven skills intelligence engine that could power a broad range of solutions including talent mobility, skills-based learning, and more. This could help companies make decisions around identified skills gaps and a number of other talent needs. It was a much bigger and pressing problem than talent mobility and a promising pivot opportunity. So Heather set about convincing the Hitch team to double down on their data science and skills intelligence differentiation and start positioning their features to address the underlying problem customers had not yet solved.

There's much more to the story of how Hitch pivoted and grew, as you'll read in the pages ahead. The result of everyone's hard work was an exciting acquisition by ServiceNow. Today, the Hitch team serves as part of ServiceNow's Employee Workflows business unit. They're delivering a new suite of talent products, powered by Hitch's skills intelligence technology, which was embedded into the ServiceNow platform for a seamless, AI-driven enterprise experience.

Mistake 6: Flailing in the Water

You put your heart and soul into launching your company. It required steadfast focus and perseverance. One day at a time, one customer at a time, one new feature at a time, you built something you can sell. Yet it's still not selling the way you need it to be. It's tempting to stay in that builder mindset, fixed on a false finish line and convinced you're just one or two more bursts of activity from triumph. You're chasing something that feels just beyond your grasp. All you need is another customer, another feature, another month to finish this regatta.

Are you mistaking effort for progress? Are you really that close? Or are you just paddling furiously into the current?

When leaders finally realize the Pivot Wave is upon them, they often make the mistake of initiating many new projects and pushing their teams hard in many directions, all in an effort to feel like they're making something happen. Yet all they end up doing is creating chaos for their team.

 The first instinct to pivot is to initiate more projects and push the teams in many directions. Yet it creates more chaos.

We talked to two founders who told us their pivot stories and how they stopped fighting the undertow.

More Features; Less Focus

A classic example of flailing is searching for "the mythical feature." You might believe you're just one feature away from landing five new customers and finally achieving good growth numbers — because that's what your prospective customers are telling you. But in reality, few prospects are looking to add another difficult conversation to their day — they have plenty internally. They aren't going to tell you your product doesn't interest them. Instead, they might be giving you weak excuses of what's missing. So you keep adding features, chasing the last one needed to make it click, making your product more complex, wasting time, and stagnating your momentum.

One founder shared a challenge of a time when they were trying to market to a customer base that was too wide. The company offered a software solution for managing documents. They were navigating a huge category; literally, every business of every size and industry had documents they needed to manage. The sales team was chasing after customers this way and that way. "We didn't have a target," the founder recalled.

When a potential customer called, they'd start selling to several different departments in the company all at once. Theoretically, any department could use their product. It was a "horizontal" product — it worked for any market and any customer. The founder told us,

> "In some cases. when we had a really great meeting. we would start with that department and then they would invite some other departments. They would love us. and we would get excited. And you know what happened? We lost the deal. Because no one could agree who was going to pay for it."

The company was losing to their competitors. To address this, they furiously added features they thought would help them stand out. But customers in different industries wanted different features. Who did they want to stand out to?

"Here's the punch line," the founder recalled. We hired a marketing expert who analyzed the firm's customer base. "You'd think we would have done that," the founder said. "But no, they did it." The marketing consultant discovered that 30 to 40 percent of the company's customers were integrating their product with one highly successful software platform. This market insight was what the company needed to stop flailing and understand how to pivot. They stopped trying to sell to everyone and anyone, and instead started targeting customers who already had that software platform installed and in use. Because the platform was well-adopted, the potential customer base was still large enough to drive a successful business.

Founders often fail to realize that being all things to all people spreads you too thin. "I was probably the company's worst enemy in this respect," the founder reflected. With a clearer target, they were able to start a pivot and unstick their boat from the mud, eventually unlocking 40 percent year-over-year growth.

> Founders often fail to realize that being all things to all people spreads you too thin.

Thrashing or Thriving

In an attempt to get unstuck from not finding product-market fit, flailing might look like the feature chasing we just described, but it can also look like starting a flurry of new projects in an effort to find something that sticks. Bob Tinker was CEO of MobileIron, and he went from founding the company to its IPO in six years. Along the way, he navigated a significant pivot around 2008 when the firm stopped marketing to their existing customers to focus on meeting the needs of an emerging new market of iPhone users in the enterprise space.

MobileIron was founded to solve security management for Blackberry, Symbian, and Windows phones. But when the iPhone hit the market, their customers were yelling at them to figure out how to help them securely use their iPhones at work. Bob recalled his sales team telling him, "I know the market analysis doesn't show it being a big thing, but there's a real pain here."

Bob was facing one of the indicators listed in the introduction of this wave that it was time to pivot: a new technology or process launched that rapidly reduced or eliminated the niche his company served. "At first," Bob explained, "the pivot wasn't for survival, it was for thrival." Thrival is a concept in Bob and Tae Hea Nahm's must-read book series, *Survival to Thrival: Building the Enterprise Startup*, which refers to not simply surviving challenges but flourishing in the midst of them.

Market data suggested the safer path was to stay the course with the support of Blackberry and Windows, since those products still had strong market share. But that's not what he did. "Market analysis can tell you a lot, but market analysis is backward-looking. Pain is forward-looking," Bob said. "We went whole hog after helping IT organizations say 'yes' to the iPhone."

> "Market analysis can tell you a lot, but market analysis is backward-looking. Pain is forward-looking." — Bob Tinker

Sometimes initiating new projects as part of a pivot isn't necessarily flailing, but it can feel like it for your employees. "When you're the CEO of a company, the good news is you control the direction of the company. So you can slam the rudder. But you also have to pay attention to the temperature of your team," Bob said. "The team is working on A, B, and C because, you know, 90 days ago you told them it was really important, right? When a leader tells their team, 'Hey, that's not that important anymore. There's this other thing,' they have to be very mindful of not thrashing the team," Bob warned. "It's important to recognize whether it's a thrash or whether you're slamming the rudder for survival or thrival. Sometimes it's hard to tell the difference."

Bob decided to start the pivot by setting up a small skunk-works operation behind the scenes. He and the product person sketched out an app and had someone else on the team build it. "We had a sales rep who was pretty interested in talking to customers about it, so we just sort of did it on the side. And then pretty quickly, a lot of people in the company were like, 'Hey, tell me more.' All of a sudden, everybody thought it was a good idea," Bob said.

> Side Note: Skunkworks is a term for using a small group outside of usual research and development to develop a project. It's named for the aircraft manufacturer Skunk Works, who used a secret R&D team. Sometimes we see the skunkworks team grow to so many employees you might as well have told your entire company, defeating the purpose of this group. This is an indication of a team scaling mistake, which we'll discuss in The Third Wave: Scaling Rough Seas.

Focus, Don't Flail

Pivots, by definition, require you to change course. Leaders who feel the pressure to pivot sometimes initiate project after project, hoping that one will "stick" and solve their problems. Yet if you change course too often, your team will feel pulled in too many directions and lose focus. And you could create a wave of panic from the dread of having evermore new ideas thrown at them. Oftentimes, the right thing to do is less, rather than more. Both stories in this chapter underscore the value of focusing as a pivot strategy.

Oftentimes, the right thing to do is less, rather than more. Successful pivots require focus.

In the first story, the company realized it didn't need more features — it needed to narrow its target audience from "all companies" to "companies who use software platform X." When the business isn't going well, it's natural for leaders to start adding more features in the hopes of satisfying every potential customer request. This approach makes your solution more complex and harder to adopt and maintain, and it also wears down your team. Leaders do this when they ignore the indicators that a pivot is necessary — just "one more feature" will put us back on track. If you see your team doing this, or if you see yourself doing it, take a step back and assess whether or not you should consider pivoting some part of your business more intentionally.

In the second story, MobileIron was willing to cannibalize its existing business to focus on the emerging and promising iPhone users (but not without having a small test first). When you aren't sure which direction you're considering pivoting, don't bring in a large portion of your company to test out uncertain changes too early. Bob's team avoided the flail by starting small and discreetly, bringing his new idea to a small group to assess until he was sure he had something that inspired others to jump aboard. By quickly taking a small first step, he could then jump in "whole hog" with his team right alongside him, instead of trying to drive a new product while continuing to grow their existing one.

Your leadership and vision will be needed more than ever to avoid spinning in circles and, instead, come to terms with a more considered approach to overcoming your Pivot Wave.

Key Takeaways

1 Trying to pivot in multiple directions at the same time will create confusion and more flailing among your team. Successful pivots often result from more focus rather than more options.

2 Beware of the "mythical feature" trap. Are you avoiding a pivot by placing all your hopes on more features that might eventually make your customers happier?

3 If you know you need to pivot but you don't yet know in which direction, begin assessing your options with a small skunkworks team.

Focus on Being Data-Informed, not Data-Driven

Most leaders would like to think they're making data-driven decisions. They can look to the past, analyze it, and use that information to drive their future choices. It's a valuable approach in many contexts, but making decisions this way can lead to a loop of incrementalism that doesn't move the needle as far or as fast as is needed for a timely pivot. Being data-driven is comfortable. It's easy to let the numbers do the job for you, as opposed to thinking through decisions in a greater context.

In a pivot, we like to aim for being data-informed — which means that you've considered the past (data), but you're using it in the context of a lot of other factors that might not have manifested in the data yet (signals). Being data-informed allows startups to strike a balance between data analysis and entrepreneurial instincts, which can be crucial for their growth and success for several reasons:

> Limited Data Available: In the early stages of a startup, there may be limited data available, making it challenging to make data-driven decisions. Being data-informed acknowledges this limitation and encourages startups to look for signals and make decisions based on the best available information.

> Flexibility and Speed: Startups often need to adapt quickly to changing market conditions, customer preferences, and technological advancements. Being data-informed allows for flexibility in decision-making, enabling startups to explore creative ideas and solutions that aren't often immediately supported by historical data. Limiting decisions by focusing only on data about what has worked in the past can stifle innovation.

> Resource Constraints: Startups typically have limited resources in terms of time, people, and money, so collecting and analyzing extensive data can be resource intensive. Being data-informed encourages startups to focus on essential data points and use them wisely to inform their decisions.

Mistake 7: Too Attached to Pivot

Some leaders are too attached — too attached to their current product, current marketing positioning, or the story they've built up around their business. It makes sense. By the time you've made it through the Launch Wave, you've spent most of your time selling your company's story and why it's uniquely positioned to succeed. It can be very hard to turn around and say, "Wait a minute. What we really need to be doing is something else."

In preparing you for your Launch Wave, we talked about ego, confirmation bias, and all the reasons it's hard to detach yourself from your beliefs and assumptions. When you're facing the Pivot Wave, detaching becomes existential. The longer you keep believing the stories you've created, the more likely you are to run out of time and money. It always takes longer to close a sale, secure an investment, or master the Pivot Wave than you think it will.

> When you're facing the Pivot Wave, detaching becomes existential.

Detaching is hard because you've put in so much of the proverbial blood, sweat, and tears in creating and proselytizing what you believed was such a good story. You needed to attract a team and funding. To do that, you built a compelling story about your ambitious opportunity and why now was the time to make it happen. You pitched your business plan with passion and conviction. But if your customers, your market size, your internal execution, or your bank account are showing signs you need to pivot, then you'll need to let go of the ideas and beliefs you've been holding on to.

It always takes longer to close a sale, secure
an investment, or master the Pivot Wave than
you think it will.

We've seen companies get themselves unstuck from
past beliefs by taking two big steps: first, by letting go of the
assumptions they were dragging along with them; and second, by
finding inspiration and being open and willing to consider other
possible futures.

Let Go of Your Story and Your Assumptions

The first step in letting go of your story is preparing yourself
to let go of the assumptions you made during your Launch Wave.
When you started this adventure, you made a lot of assumptions
that drove your decisions around your target customers, the
problem you were solving for, and the solution you were building.
If you need to pivot, then some of those assumptions were incorrect
— so it's time to examine them all over again and determine which
ones are weighing your startup down.

Sometimes getting outside help for this step is crucial.
As much as you believe in yourself, it's far more likely you need
someone with fresh eyes to help you identify and throw overboard
the jetsam that's weighing your company down. One of the most
common characteristics of entrepreneurs is overconfidence — a
necessary trait when taking on the risk of a new venture. However,
if you find yourself facing a Pivot Wave, overconfidence may hinder
your ability to see the warning indicators. Usually, the closer
you are to the original idea, the harder it is to pivot from it. In a
way, a certain part of your identity is attached to the idea you've
developed, and it's hard for you to let that go. But if you can't let
go of your assumptions and reposition your business, your board
might do the repositioning for you. And you might find yourself

replaced by someone who doesn't have the baggage of previous assumptions to hold onto.

To help you objectively assess your situation, we ask you to stop being — for a few days — the unwavering, confident entrepreneur and schedule an offsite meeting with your team or board of advisors to examine where things stand. Think about your target customer or audience, the problem you're solving, and the solution you've built. What's working? What's not? You would need to evaluate your solution in different scenarios.

> Side Note: Solutions could be products, and most people think of the word this way. But we intend to use "solutions" in a much broader capacity to cover not just what people consider "products" a company might sell. In search of finding product-market fit, entrepreneurs often think about pivoting by changing the "P" part of the equation — the product. But a solution may be any number of things: a service offered, the way a company packages its offerings, the method the company employs to acquire customers, a company's support or maintenance offerings, or even its business model. Each one of these are solutions businesses employ in order to grow.

To remind entrepreneurs that pivots are often about changing something other than their core product, like the audience they target, or how the whole solution works for that market, we use the term "business-market fit" when advising on pivot strategies.

> Business-market fit is a holistic approach to evaluating the degree to which you have strong market demand — looking at your product or solution, people, messaging, positioning, and approach to the market — and demonstrating you have a unique business offering that people desperately want.

We've put together a simple Assumptions Assessment you can use to get started evaluating your situation.

Assumptions Assessment

Scenario 1

✓ Target
Audience

✓ Problem to
Be Solved   Take a shot of tequila!

✓ Solution
Built

Scenario 2

✓ Target
Audience

✓ Problem to
Be Solved

✗ Solution
Built

Enough people have the problem you've identified and they're telling you they would pay you to solve it. But they aren't willing to pay for what you've built. Do you have a user experience issue? A functionality issue? A pricing issue? You'll have to test a new set of assumptions to decide if you should continue iterating on your MVP or if you need to build a new MVP.

Scenario 3

✓ Target
Audience

✗ Problem to
Be Solved

✗ Solution
Built

This is a target audience you want to keep. Either you're an expert on it, have built relationships in it, or have unique insights about it that give you a competitive advantage. But you're probably only solving a nice-to-have problem (at best). You'll need to test a new set of assumptions to find a bigger pain point to solve and a new MVP. Reuse pieces of your existing solution if possible. Consider your runway.

Scenario 4

✓ Target
Audience

✗ Problem to
Be Solved

✓ Solution
Built

Customers are using your solution for something you didn't intend. Figure out why. Figure out if the market for this new problem is large enough to support the business you want to build. Once you've figured that out, build a company around it. (That's how Discord pivoted from building a game to a communications tool for gamers: same audience, good product (part of it), but they were solving something entirely different.[1])

Scenario 5

✗ Target Audience

✓ Problem to Be Solved

✗ Solution Built

You're likely passionate about the domain and the problem space. But your solution isn't speaking to any audience you've tried it with. Start talking to new target audience segments to see if the problem you've identified is a real pain point for them. Also examine your assumptions around your pricing model and how you go to market.

Scenario 6

✗ Target Audience

✓ Problem to Be Solved

✓ Solution Built

Someone loves to use your solution to solve their problem, but it's not the audience you expected. You'll only discover this if you start analyzing usage by segment. You need to figure out if there is an audience/market big enough to sustain your business.

Scenario 7

✗ Target Audience

✗ Problem to Be Solved

✓ Solution Built

You have a solution that's not quite working for the problem you intended to solve or for the customers you intended to sell to. But you're convinced that your solution — usually one that's technically differentiated enough — must be good for something or someone. This is the typical case of a technology searching for a problem to fix. Start over with a new set of assumptions.

Scenario 8

✗ Target Audience

✗ Problem to Be Solved

✗ Solution Built

Take a shot of tequila!

---

1 Coogan, "How Jason," John Coogan.

Pivoting the Target Audience

A B2B SaaS company we know hired a new VP of Sales because it hadn't been hitting its sales targets. The VP was experienced in selling into the SMB market and quickly assembled a large sales and marketing team with the same focus.

Despite this investment, the company still wasn't achieving its growth targets, and now the new team was quickly draining the company's cash reserves. That's when Heather was brought on to assess the situation. Her analysis showed the company was generating 85 percent of its revenue from 50 percent of its customers — and that 50 percent were large enterprises, not SMBs. The smaller accounts were easier to close, but they sucked up a disproportionate amount of internal resources for the 15 percent of revenue they were contributing to the company.

The needed pivot wasn't about building a different product or solving a different problem; it needed to pivot the target audience — Scenario 6 in our Assumptions Assessment.

The hard part was helping the CEO and board let go of their assumptions. Neither wanted to let go of their SMB customers. They'd already invested so much in acquiring them. The revenue they generated, while not a significant part of the company's overall revenue, was still cash the company badly needed. But in assuming they couldn't part with this market segment, they were failing to consider the opportunity costs of continuing to support it. The company was spending a lot on a sales and marketing team and paid marketing strategy that wasn't aligned with the pivot it needed to make. Testing this theory, Heather turned off their paid marketing channels, to the horror of the team, only to find that . . . nothing bad happened. The paid marketing was focused on the SMB market — the equivalent of throwing $25,000 a month in the trash, when that money could be deployed in support of an account-based marketing strategy to acquire large enterprise customers.

To convince the CEO and board to change direction, Heather had to show them a better opportunity existed. She had to find The

Crack in the Market. In the end, the company did make the needed pivot and was able to unlock growth that had been stalled for many years. Annual recurring revenue grew from $4 million to more than $8 million in less than 18 months, and the company went from losing $2 million per year to profitability.

Finding Inspiration: The Crack in the Market

Many leaders don't make it to this point. It's easy to act quickly out of fear and desperation — to start trying new things without a methodical approach. But if you can let go of the assumptions that have been holding you back and assess what needs to change, then you'll be ready to start exploring where you have to pivot. Just like during your Launch Wave, you have to go back to listening — to the right people. Unlike your Launch Wave, you aren't starting from a blank slate. If you're pivoting (not just giving up and starting a new business) then you're looking for a new approach that leverages some of what you've already developed — the "good" elements from your Assumptions Assessment. These elements get you to a new business opportunity faster, but they also act as constraints on what you can pursue.

Finding an opportunity within those constraints is what we call finding The Crack in the Market. The Crack in the Market is a small opening that presents itself when you've listened to the indicator warning that something is wrong and you're able to look, with fresh eyes, at the signals (data, feedback, etc.) from all around you that reveal new opportunities to harness the value of your capabilities. To find The Crack in the Market, you need to gather information about your business, understand the strengths you have built, and see how they match up with the opportunities in the market.

Finding The Crack in the Market starts with listening to the indicator warning that something is wrong, alerting you to pay attention to signals in the market.

Heather found The Crack in the Market for Hitch's business by doing three things. First, she undertook a listening tour with Hitch's existing and prospective customers, asking open-ended questions that allowed the conversation to flow to where the customer pain points were. Second, she took the signals she heard from her listening tour and then took a closer look at the existing Hitch team and technology. From her initial assessment of the business, she knew Hitch had a solution that some customers liked. But she needed it pointed at a different problem, and maybe a different customer type, following Scenario 7 in our Assumptions Assessment. Third, knowing what she needed to pivot, she focused on other problems identified during her listening tour and market evaluation to determine what could be solved by harnessing Hitch's proprietary algorithms, and ultimately repositioning it as a dynamic AI-driven, skills intelligence platform.

How Heather moved from finding The Crack in the Market to making the pivot in practice is covered in the next chapter. Before we get there, we have another founder story that illustrates many of the pivot steps we've discussed so far.

From Nuclear Power Plants to Bakeries — Find Problems You Can Solve

Taiga Robotics is a pivot story within a pivot story. Like many successful founders, Dmitri Ignakov and Ilija Jovanovic met in grad school. Unlike most founders, they were doing research in aerospace engineering — they were literally rocket scientists. It was during that time the Fukushima Daiichi nuclear power plant disaster occurred. The clean-up efforts were beyond challenging. Highly radioactive nuclear material had to be collected and placed in shielded containers. People looked to robots to do the work, but it was too hard to program them to pick through all the debris and automate their activity in such an unstructured environment.

Dmitri and Ilija set out to find a solution. The challenge was that programming an autonomous robot with the level of dexterity, flexibility, and intelligence to navigate unstructured environments was a multi-tens of millions of dollars prospect and a multi-year project. They had deep experience in robotic motion controls and AI. Their "aha" moment was connecting a robot's control system to a human through a virtual reality (VR) platform, allowing a human to remotely maneuver an intelligent robot from a safe distance.

Dmitri and Ilija founded Taiga Robotics. Their pitch won first prize at a competition sponsored by Ontario Power Generation (OPG). It set them up to work with OPG as their first paying customer, helping the OPG-owned Darlington Nuclear Generating Station with its refurbishment efforts.

Within two years, they were running six-figure pilot projects with OPG and securing pilot projects with customers in other large industrial segments in which safety was an issue. "At the time, it felt like a lot of money was coming in. The check sizes were getting bigger, and the customers were staying engaged," said Dmitri.

To the credit of the founders, even though they were generating revenue, they decided to take a step back and examine the direction in which they were sailing. And through this, they realized two things. "The projects we were getting were mostly pilot projects or small orders from the 'innovation' groups within large companies." Ilija said. "The projects were large, drawn-out affairs with long deployment cycles."

While they were selling to individual innovators within the definition of Geoffrey Moore's *Crossing the Chasm*, they were innovators within organizations that were, at best, early majority, and more likely, late majority companies. The innovators had some budget, but most didn't have the authority to make real product decisions on behalf of a whole business. The team realized they couldn't launch a startup selling first to the early and late majority. They had to pivot their product for a different customer. (Go to the end of **Mistake 7: Too Attached to Pivot** to read the Sidebar: The Rigth Customers for Each Wave.)

It's important to understand the type of company you're selling to, and more importantly, what is the decision-making power of the group, department, or business you're selling into.

The second realization was the market they were after was too small when they considered their total addressable market (TAM), service addressable market (SAM), and their serviceable obtainable market (SOM).

TAM SAM SOM

TOTAL ADDRESSABLE MARKET (TAM)
THE TOTAL POTENTIAL MARKET DEMAND FOR A PRODUCT OR SERVICE, ASSUMING 100% OF THE MARKET IS CAPTURED.

TAM
SAM
SOM

SERVICEABLE ADDRESSABLE MARKET (SAM)
THE PART OF TAM THAT CAN ACTUALLY BE REACHED.

SERVICEABLE OBTAINABLE MARKET (SOM)
AN ESTIMATE OF THE PORTION OF REVENUE WITHIN A SPECIFIC PRODUCT SEGMENT A COMPANY IS ABLE TO CAPTURE.

There were only so many of these large industrial companies in the United States and Canada, which was their first target market. Of those, many were unionized, which would make it hard for the companies to move work away from humans to robots, even though some humans would still be needed to manage the robots. Additionally, the players in this market segment weren't early adopters of technological change. The market segment they'd targeted, while it looked attractive at the start, suddenly didn't feel large enough.

Dmitri said it was around that time he heard from a friend operating a bakery. The year was 2020, COVID-19 was devastating communities, and she wasn't able to operate her bakery. She offhandedly chided,

---

"You can make a robot that moves nuclear material to shielded containers. Why can't you get me a robot to help me make a cake?"

The question was a joke, but Dmitri saw it as a challenge and an opportunity to explore a pivot. This would be a pivot away from industrial customers to commercial customers. It would also require a product pivot, following Scenario 5 in our Assumptions Assessment. Commercial customers didn't need the VR unit — they weren't operating in dangerous conditions. What they needed was Taiga's underlying robotics control software and AI training models.

Luckily for the team, an industrial customer had already asked them to pitch just the software part of their platform. They had noticed in that case the sales cycle had moved much faster. That was the first indication that a pivot was worth investigating.

What was their software competitive advantage? First, their software was designed to operate in a variety of environments. It could maneuver robots — really any digital machine — through nuclear debris, in mining shafts, in kitchens, warehouses, or greenhouses. Most other robotic systems were purpose-built for a specific environment, but Taiga had decoupled the intelligence from the machine form, so any robotic system could take advantage of their software. Second, Taiga's AI training algorithms allowed the software to train the robotic system onsite without technical assistance. This allowed one Taiga robot to be trained to do multiple tasks when they were needed. Perhaps the robot might pack boxes during the day and shelve boxes during the evening. Finally, because the software could work with any machine design, they could design the hardware part of the system to work in confined spaces. To be sure, there was still plenty of development work to be done to pivot the product, but the core technology was all there from the start.

As the Taiga team changed their perspective from industrial verticals to commercial verticals — and then even further, to how their technology differentiators matched specific segments within a commercial vertical — they started recognizing signals. The signals showed up in the kinds of customers that had problems they could solve and ones they could actually sell to.

 Looking at the signals in the market, you need to gather information about your business, understand the strengths you've built, and see how they match up with the opportunities in the market. In some cases, you may need to overcome resistance to the pivot to go all-in.

Today, Taiga Robotics is working with commercial companies to automate tasks by using their pick-and-place robot, PTeR. These robots are trained by AI to work in small industrial operations to do everything from packaging to sorting to kitting products. They can even do industrial bakery automation, as Dmitri's friend inspired. The Taiga Robotics team successfully pivoted and has made it simple for manufacturing and processing SMBs to get up and running with a robot through an easy three-step process and low monthly fee.

How to Find The Crack in the Market

Taiga's pivot was successful because they were willing to let go of the assumptions on which they'd launched. As they were navigating their Launch Wave, they found immediate interest from the industries they were targeting. With six figure deals coming in, it would be hard for any entrepreneur to pivot from that business model. To their credit, they paid attention to two indicators we identified at the start of this section:

1. All their customers were running pilot projects out of their innovation groups. The sales cycle to get out of the innovation group (and into the operations team with actual product purchasing power) was too long for the startup to support.

2. These customers loved their product. But even if they did get the product sold within the operations groups, it wasn't clear the overall size of the market for this product was sufficient to support the growth they wanted to see.

These were important insights for the Taiga team to uncover. It showed them it was time to pivot. But it didn't help them understand where to pivot to. For a while, they were still trying to apply their existing VR-interfaced robotics solution to other market segments. It wasn't until they started paying attention to the indicator warnings, listening to the signals and what others were telling them — including their bakery owner friend — that they began seeing The Crack in the Market.

Like Taiga and Hitch, you can find your Crack in the Market by following the same steps.

GATHER BUSINESS
INFORMATION

TAKE INVENTORY OF
YOUR STRENGTHS

CONNECT THE
DOTS

Step 1: Gather Business Information

Your goal in this step is to take a fresh look at the reality of the market, your business, and your product — with an emphasis on fresh. As you've seen in this chapter, it all starts with letting go of your assumptions; only when you do that will you be ready to listen with that fresh perspective.

Practically, this means going into a new research phase. Talk to your customers. Talk to people who aren't your customers but who might have a problem your current solution can be modified to solve. Talk to different types of users of your solution — are they using it in surprising ways? Talk to analysts and thought leaders about your market and adjacencies. Find out what they think are the current trends your customers are paying attention to these days.

And while you'll need your entire team to gather these insights, it's important that you hear them directly. Neither the Taiga founders nor Heather at Hitch delegated this step solely to their teams. They were having exploratory conversations with their market segments directly. As with the Launch Wave, hearing directly from customers, potential customers, and non-customers removes any filter and bias that could be introduced through your customer support or sales teams.

Step 2: Take Inventory of Your Strengths

It's now abundantly clear that in a pivot you're not starting from scratch. You have insights, customers, a team, and even a product. You're still evaluating what can be reused as you plot your new direction, so it's important to account for all of the strengths you already have that can be deployed to maneuver this Crack in the Market.

> ❯ Your strengths could be in any part of the solution you've built. It might be hidden deep within your organization, like Hitch using their AI algorithms for talent mobility, or for some adjacent function, like the communication stack Discord built as part of their original game.

> ❯ Or it might be in the team you've built. If you built the wrong solution, but your team has the right expertise, you can leverage your team's expertise to build the right solution. For example, if you've been pursuing a direct sales strategy with a weak sales team, but you have an amazing marketing team, then perhaps your Crack in the Market is to change your business to a marketing-led model.

> ❯ Or it may be the relationships and deep insights you've built with certain companies or parts of an industry.

> ❯ Or something else . . .

Step 3: Connect the Dots

Market signals on one side, strengths on the other side: with this information laid out, start developing a new vision of the future, some new hypotheses, and how they match up with the opportunities in the market. It's both an art and a science to find The Crack in the Market and a new direction for your company.

Taiga's market signals were: they were having a hard time getting out of the innovation teams of the first market segment they targeted, and a customer had approached them about licensing a narrower part of their product for commercial use. They connected these market signals with a re-examination of their strengths — their underlying robotics software was highly adaptable to different environments and easily trained for different tasks. This led them to pivot to selling a revised version of their product to a different target audience that didn't need years of pilot projects before making a production decision.

Similarly, Hitch's market signals indicated they hadn't been solving the real pain point their customers and prospects were feeling. Understanding that pain point allowed them to connect the dots to the internal strengths around AI and skills intelligence data that could be positioned to address this pain point. This connection drove their pivot strategy of product repositioning, combined with a lot of market repositioning.

Neither of these pivots required a complete product overhaul to be able to take advantage of The Crack in the Market. This is a great situation in that it enables a faster pivot than having to develop a whole new offering. Other possibly more agile pivots might involve:

› Offering your same product under a different pricing strategy that fits your target audience's buying behavior better.

› Packaging product features to allow different users to take advantage of different benefits without having to pay for the full offering.

› Selling to a different target audience that has a greater need for the product you're offering.

› Adding post-sales customer support that helps your existing customers make full use of your existing product, so they don't churn.

The only way you'll know which direction you should go is by connecting the market signals to your existing strengths.

Key Takeaways

---

1 When you see indicators that you need to pivot, take the time to look at your business, let go of your assumptions and the story you've been telling everyone, and work with a team to complete our Assumptions Assessment or a similar framework. Remember, don't do this alone.

---

2 To find The Crack in the Market, you need to gather business information, take inventory of your strengths, and connect the dots between the market signals you're getting and the strengths you can leverage. You'll need to create a new set of assumptions and validate them.

---

3 Pivots aren't just about changing your product — you need to consider business-market fit. Before you rush to build a new offering, consider whether your market signals are telling you that you might need to readjust something else instead — perhaps your marketing strategy, how you price or package your product, who you're selling to, or even how you support your customers.

Sidebar:
The Right Customers for Each Wave

Geoffrey Moore's *Crossing the Chasm: Marketing and Selling Disruptive Products to Mainstream Customers*, first published in 1991, long before even the first internet wave, has withstood the test of time. If you're building a business and haven't read it, finish *Sail to Scale* first, and then read his next.

Moore groups customers into five psychographic categories: innovators, early adopters, early majority, late majority, and laggards. Understanding the characteristics of each customer type is critical for founders trying to navigate the Four Waves of startup growth. You'll waste precious time if you're navigating a Launch or Pivot Wave and find yourself trying to sell to a customer in an early or late majority category.

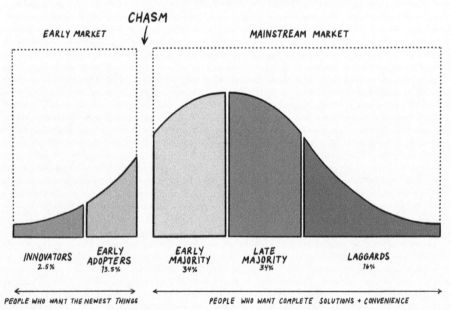

Source: Geoffrey Moore, *Crossing the Chasm*

To help you better connect the customers you should target against the waves you're currently facing, we've created this quick guide with the characteristics of customers in each category, how to identify them, and how to keep them engaged if you managed to sell them your product.

You can also use this guide throughout each of your Four Waves, to make sure you aren't trying to target a customer too far down the market curve.

Innovators

Three Characteristics of an Innovator

1. Creative
2. Curious
3. Risk-taker

How to Identify the Innovator

› Easy to engage with in a conversation about a new product offering
› If the new innovation interests them, they'll ask lots of questions (and if it doesn't, then it'll be hard to identify them because they won't)
› Easy to set up a follow up meeting or schedule a demo
› They don't ask about price immediately

How to Keep the Innovator Engaged Post-Sale

› Add them to your community because they're likely to engage in discussions as your community grows.
› Give them access to a solutions engineer or product manager (depending on how you define these roles in your company); innovators are tech enthusiasts and can provide lots of (too many) ideas to your team.

Early Adopters

Three Characteristics of an Early Adopter

1. Long-term thinker
2. Skilled communicator
3. Strategic risk-taker

How to Identify the Early Adopter

> Talks or asks about the possibilities of where your product can go (considers the future state as well as the present state)
> Almost helps you understand the potential of your own product
> Might think about other people who could be interested in what you're offering
> Unlike innovators who might be all over the place in your conversation, early adopters tend to be much more focused on how your offering can address their problems today and in the future

How to Keep the Early Adopter Engaged Post-Sale

> Keep them regularly informed of new product updates, market dynamics, and trends.
> Give them visibility into your product roadmap.

Early Majority

Three Characteristics of the Early Majority

1. Results-oriented
2. Rational
3. Bottom-line oriented

How to Identify the Early Majority

> They won't engage with you if you talk in the abstract; to connect with them, talk specifics
> They ask questions about how your product works — and what they really want to know (whether they say it out loud or not) is how it works in their workflow and their daily routines
> They ask you about pricing early in the conversation
> It's difficult to get more time with them; if you get another meeting, it'll likely be weeks or more away

How to Keep the Early Majority Engaged Post-Sale

> Create efficient means of communicating only what they need to know to succeed in integrating your product into their environment effectively.
> Make sure your product or your account manager can show ROI on product usage.

Late Majority

Three Characteristics of the Late Majority

1. Belief in what is time-tested
2. Hierarchical
3. Prudent restraint

How to Identify the Late Majority

> They're interested in references from established customers
> They'll ask questions about how much money they can save by using your product
> They might leave pricing negotiations to finance or another team
> They'll demand all necessary certifications and integrations

How to Keep the Late Majority Engaged Post-Sale

> Integrate quickly into their workflow and environment; have professional services ready and prepared to deliver effectively.
> Deliver exceptional customer support.

Laggards

Three Characteristics of a Laggard

1. Mistrustful
2. Pessimistic
3. Evidence-based

How to Identify the Laggard

> Similar to the late majority, they'll be focused on cost, certifications, and integrations
> They might request service-level agreements in your contract beyond what's standard
> Outbound sales rarely works — you're more likely to connect with them when they decide to reach out to you
> They're likely using products in their business from two generations ago

How to Keep the Laggard Engaged Post-Sale

> Integrate quickly into their workflow and environment; have professional services ready and prepared to deliver effectively.
> Don't ask them to upgrade to new versions very often; support their current product version for a long time.

Mistake 8: Ignoring False Commitment

As a team member, you may have been hit by a poorly executed Pivot Wave before. Your leadership shows up one day with a "brand new strategy." It's unclear what should happen with the ongoing work. It's unclear how this new strategy will become a reality. And it's not the first time it's happened either — this is just the new flavor of the quarter. So, knowing that "this too shall pass," you nod and carry on half-heartedly with whatever you're being asked to do to make this "bold" new direction a reality. In the meantime, your trust and faith in this company diminishes another notch, and you update your LinkedIn profile.

Every pivot of any significance is a change management effort. Most leaders focus primarily on the execution of the pivot as seen outside their walls (the product, the messaging, the business model, etc.) and fail to manage the change that needs to happen in the building with their team. They speed through the Pivot Wave without realizing their crew is falling behind because they haven't really committed to the pivot. This mistake happens in companies of all sizes, but in a startup, where bandwidth is scarce, you can't afford to have a lost team for any period of time. You need everybody on the new course ASAP.

> Every pivot of any significance is a change management effort. Most leaders focus primarily on the execution of the pivot as seen outside their walls and fail to manage the change that needs to happen in the building with their team.

There have been many excellent books published about managing through change. One of the classics is *Leading Change* by John P. Kotter, along with his subsequent change management work.[3] To successfully execute a pivot, we focus on recognizing false commitment and the change management effort involved in overcoming people's natural resistance to change.

Recognizing False Commitment

If you don't spend time assessing and addressing whether you have the team behind you in the pivot you're about to make, then you aren't likely to keep your ship afloat. Here are some signs that you have either no commitment or false commitment from your team:

> You put forth your bold new plan, and no one has any questions.

> You find yourself always having to follow up on projects — there doesn't seem to be any sense of urgency.

> Things people keep saying they're going to fix don't seem to get fixed.

> You start hearing employees complaining about a lack of strategy, which is odd because you believe you've been very clear about the new strategy.

> There's animated conversation, until you walk in the room.

People are naturally resistant to change because it creates FUD (fear, uncertainty, and doubt). The fears could be related to the new direction not matching their skills, disrupting their routines, or having to take on more work. The uncertainty could be related to mistrust in the viability of the new direction and the lack of control over change imposed on them. The doubts could be related to the team being too attached to the previous direction or not understanding the new one well enough.

---

3 Kotter International.

Internal FUD during a pivot can be deadly for your business and must be managed the same way the pivot is managed for external stakeholders. If you hired entrepreneur archetype team members (see **Archetypes and the Entrepreneur Archetype** in **Mistake 2: Building the Wrong Minimum Viable Team**), the FUD will be more palatable to them, as this archetype is less risk averse and more comfortable with ambiguity and change. Still, it needs to be managed.

Overcoming Resistance to Change

You can find a lot on the internet about how to build a more committed team. Recognize their work. Reward them with incentives and bonuses. Deliver training and development. Create clear end goals.

Perhaps these make sense for a larger, more established company sailing through calmer waters. However, during a pivot, you're in rough seas facing a wall of water coming at you. This isn't the time for cheerleading. Like every wave you face, this requires real leadership — and real, fast change management practices.

> Like every wave you face, this requires real leadership — and real, fast change management practices.

To manage your team through a change, we like the Satir Change Model, a framework developed by Virginia Satir.[4] She was a family therapist, but her five-stage model has been used in organizational change management. Unlike other change management frameworks, this model focuses on the human element of change, so it's directly relevant to addressing false commitment.

---

4 Satir, *The Satir Model*

The five changes are:

1. Late Status Quo: The current state of familiar ways of doing things

2. The Foreign Element and Resulting Resistance: The indication of a need for a pivot that reveals problems with critical assumptions participants refuse to heed.

3. Chaos: As the change begins to be implemented, things feel in disarray and people feel overwhelmed and uncertain.

4. Integration: Changes become integrated into participants' daily routines, and a sense of stability and comfort takes shape.

5. New Status Quo: The change becomes the new norm, and participants no longer consider it a "new way of doing things."

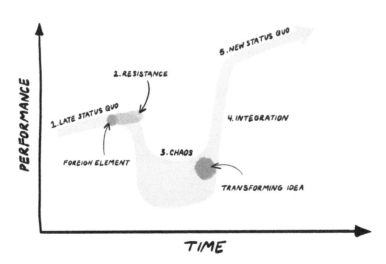

Source: Virginia Satir, The Satir Model

During a pivot, you'll inevitably go through the chaos stage, where FUD is at a maximum, and nobody really knows how to move from the old to the new yet. To move from chaos to integrating the pivot into your company's new way of doing things, you need to deliver a Transforming Idea.

The Transforming Idea

A similar period of chaos happened during the Hitch pivot. Some members of the team strongly resisted the change. Heather's first step was to frame the new positioning — skills intelligence versus talent mobility — as the Transforming Idea. Essentially she bottled all the data she'd collected with all the inspiration she felt after connecting the dots between the market signals and the company's strengths, and brought it to the rest of the company — and beyond to their customers, investors, and market in general. Had Heather found The Crack in the Market for Hitch and just started implementing the change, she might not have been able to bring all the other stakeholders along, and the Hitch pivot could've failed.

A pivot strategy only becomes a Transforming Idea when you overcome your stakeholders' resistance to it. Doing this requires making the change exciting enough that people don't want to be left behind. Hitch immediately changed its internal and external messaging to lead as a data-first, skills intelligence platform for the future of work. This drove enthusiasm inside and outside the company. Press releases emphasized the value the change brought to Hitch's customers. Hitch's dynamic skills graph captured the attention of industry analysts and market luminaries, who also got behind the company. It's this purposeful effort to drive enthusiasm that turns a great idea into a Transforming Idea.

One Shot to Make It Perfect

We love the near-death story of Nvidia, the multi-trillion market cap graphics and AI platform darling of the markets, as perhaps the iconic example of a pivot and Transforming Idea. Jensen Huang, Nvidia's celebrated co-founder and CEO, is quoted in an interview with Acquired podcast as setting this context:[5]

> *"That time — 1997 — was probably Nvidia's best moment. The reason for that was our backs were up against the wall. We were running out of time, were running out of money, and for a lot of employees, running out of hope. The question is, what do we do?"*

The company had taken a direction in building its 3D graphics chips that the market didn't follow. Jensen and his team realized their Crack in the Market. The PC market was driving a new and growing audience of gaming enthusiasts who would pay anything for the fastest gaming chip they could get their hands on. So they decided to build the most outrageously powerful chip the market could imagine — "as big as physics could afford at that time." But they didn't have enough runway to fully build a test chip. So they decided to do what no one else was doing at the time. They decided to risk it and emulate their chip instead of manufacturing it completely.

Emulation is like simulating the results of using both the hardware and software of the chip together. But it's not the real thing. And they couldn't be sure they'd catch all the bugs taking this approach. Here's the best quote from that interview:

> *"I remember having a conversation with our leaders, and they said, but Jensen, how do you know it's going to be perfect? I said, I know it's going to be perfect, because if it's not, we'll be out of business. So let's make it perfect. We get one shot."*

---

5 Huang, interview.

Jensen's Transforming Idea was designing the biggest 3D graphics chip that physics could allow without fully testing it before it went to market. His confidence, daring, and the sheer audacity of the plan couldn't help but push his team through the chaos stage and into the next stage of commitment — integration.

Integrating Your Team into the Pivot

Once the chaos phase is over, your team will begin to settle into their new reality. This is a crucial time to bring the team with you and make your pivot an even better opportunity than the vision and ideas you brought to the table. Starting a pivot needs a leader. Executing a pivot needs the entire team.

> Starting a pivot needs a leader. Executing a pivot needs the entire team.

The integration phase is usually full of creativity. As the team internalizes the Transforming Idea, every team member, every department starts ideating what they can bring to the table to make it better. Here are some best practices for this time.

You Can't Repeat It Too Often

Basketball coach John Wooden said:[6]

> "I created eight laws of learning; namely, explanation, demonstration, imitation, repetition, repetition, repetition, repetition, repetition."

Remember the early days of your startup when you drove excitement and commitment by repeating your vision to everybody who would listen, about a hundred times a week? Well, you have to do it again — and probably more often — because in a pivot you

---

6 O'Reilly Media, Inc., "The Laws," O'Rielly.

have to undo the old strategy before getting them to believe in this one. Use as many forums and formats as possible to tell the story of your new direction. Create slogans. Put up posters. Hold frequent, all-hands meetings. Schedule more one-on-ones with key contributors in your company.

Harness the Energy

Once people "get it," the team will be energized by the new direction. This is the type of energy you need to sustain the change and make it to the other side. Make this pivot existential (it probably is), and make your team feel like superheroes (they probably are).

> Make this pivot existential (it probably is), and make your team feel like superheroes (they probably are).

Make the Team Part of the Solution

The best way to get buy-in is by enlisting everybody to be part of the solution. People overcome their resistance when they feel like they have agency and control over their own direction. You and your leadership team need to define the "what," but your organization should determine the "how." Make this a team activity. Organize departmental summits, hackathons, brainstorming sessions — any creative outlet that allows team members to contribute their "how" to integrate the pivot into a new status quo.

Be very clear that you are looking for hows — not new ideas — on where or what to pivot to. Otherwise, your effort to motivate your employees to integrate the pivot might backfire into increasing their frustration because you aren't listening to them.

Get the Market Behind You

Integrating the pivot into a new approach for your business isn't just about getting your team aligned. It's also about getting your external relationships to believe in your Transforming Idea. Make your external messaging audacious. Issue a press release. Speak about the change at conferences. Getting the market behind you can drive enthusiasm not only externally, but also within the company.

Embrace and Reward Change

As you let go of your assumptions and embrace a new direction, your team will also have to. You can encourage that behavior by broadly recognizing when people and teams adapt quickly to the new direction. If you have many entrepreneur archetypes in your company, then they're likely to embrace the new direction more easily and show resilience in the face of change. Enlist them to show the way. Culture is what you reward and what you punish. So, recognize and reward adaptability.

The Right Crew for the Pivot

As this is all happening, you should be doing one more thing with your executive team — writing new job descriptions for after the pivot as if you were designing the company from scratch. And then evaluate your team for their fit against these new job descriptions.

The scale of this process depends on the size of the pivot. Sometimes, it's an opportunity to let your team stretch into new skills, which can motivate them to stay on board with the new course. This is the best outcome because your dedication to the existing team will create an even stronger culture. But other times, your pivot may be targeting a completely different audience or selling a completely different business model, and the experts you need to turn the ship around aren't part of your current crew. Or there might be a few people showing signs that they might be getting seasick from the changing tides. A pivot will usually require bringing in some new expertise and energy that aligns with your new direction.

> A pivot will usually require bringing in some new expertise and energy that aligns with your new direction.

Don't forget, you're also part of the crew. The hardest thing for leaders to do is evaluate whether they need to pivot themselves into a new role. Too many times, a CEO is asked to leave because they weren't open to evaluating their own skills against the wave they were suddenly facing. Few founders are able to navigate all Four Waves successfully. And they shouldn't feel like they have to be able to do that. We all demonstrate skills in certain areas and "learning opportunities" in others. It's always best to make the decision yourself rather than have it made for you.

Key Takeaways

1	Be on the lookout for signals that you have "false commitment" from your team, and recognize that FUD (fears, uncertainty, and doubts) needs to be managed internally during a pivot.
2	Moving past the chaos to the integration of the pivot requires change management. You will need to deliver a Transforming Idea.
3	Starting a pivot needs a leader and vision. Executing a pivot needs the entire team.

Mistake 9: Not Going All-In

We didn't call this mistake "Failing to Implement Your Pivot" or "Not Following the Ten Steps to a Successful Pivot." We called it "Not Going All-In" because a pivot is a change in nearly every part of your business — it's an all-in effort. And that's not something you'll read about or perhaps even hear about from your board or advisors. Instead, the media makes pivots sound like focused changes with headlines like: *Brex started out as a VR headset company and ended up as a credit card company for startups*. Which can come across as the only thing the Brex founders had to do to succeed in their pivot was tell their product team to build something different.

Successful pivots are all-in company pivots. For an example of what happens when you don't go all in, we heard from the CEO of a B2B SaaS startup.

A Half Pivot to a New Target Audience

Having successfully navigated its Launch Wave, this B2B SaaS company was now scaling its business. Their entire Launch Wave strategy had been focused on selling their solution to product managers. That strategy got them to many hundreds of product managers using their solution across almost as many companies. Along the way, the CEO learned important data about the company's target audience: product managers don't have large budgets to spend on software and systems. Despite adding more features to their product that users loved, they weren't seeing this effort translated into sufficient revenue growth. That was the indicator that a pivot of some sort was necessary.

Obviously, selling something different to the same audience with the same budget constraints wasn't going to solve the problem. Instead, the CEO determined the company's product could solve problems experienced by marketing teams. In terms of our **Assumptions Assessment**, they identified with Scenario 6: someone loves to use your solution to solve their problem, but it's not the audience you're currently selling to.

The CEO made an effort to bring the employees along and help them overcome their resistance to change. The CEO held all-hands meetings and exuberantly introduced the new direction that was going to change the trajectory of the company. The CEO was convincing with their story and got the employees rallied and excited for the promise of renewed success.

The CEO and their leadership team then went to work on the pivot. They repackaged their existing solution specifically for product teams and created a new package for marketing teams. The underlying product was the same, but it was positioned in two different ways in marketing materials, on the website, in sales enablement documents, and in customer calls. Nine months after the roll out of this pivot, the numbers hadn't changed much. The CEO took a beat and met with a number of new customers and lost customers to whom the company had sold or tried to sell the new package. What the CEO learned is a great illustration of the mistake of not going all-in.

> The sales team was trained to sell to product managers. Speaking the language of marketing managers was foreign to them. Even though they'd been equipped with new documentation, they hadn't been adequately trained to sell to this different audience segment.

> In the same vein, the solution itself had instructions and workflows designed for product use cases. The product team hadn't been given time to modify the solution's interface to resonate better with marketing managers.

> The support team faced a similar experience. They were used to supporting product use cases. They were inexperienced with the problems that resulted from using the solution for marketing use cases, so support for the new marketing customers suffered.

Across the company, teams were struggling to juggle the distinct needs of different target audiences, all because the company changed one aspect of its business and called that a pivot, instead of going all-in with the necessary changes across the company.

Eventually, the company unwound this initial offering and invested a few more quarters in developing and implementing a roadmap redesigning its core solution to support easy reconfiguration for different audience types. It also invested in training across marketing, sales, services, and support to understand the pain points and use cases of the new audience it was trying to make inroads to. Finally, they restructured their pricing and packaging, so it was easier for a company that wanted to buy licenses for both its product and marketing teams.

Pivoting Everyone, Everywhere, All at Once

As our previous story shows, one change in a company's focus is likely to bring about a cascade of resulting new challenges. A pivot is never just a product pivot, or a target audience pivot, or a business model pivot. To conquer your Pivot Wave, you must look across your company and consider how each team needs to pivot to successfully redirect your startup. There are many questions to consider as you build a plan to pivot everyone, everywhere, all at once.

> A pivot is never just a product pivot, or a target audience pivot, or a business model pivot. To conquer your Pivot Wave, you must look across your company and consider how each team needs to pivot to successfully redirect your startup.

Solving a Different Problem

If you're pivoting to solve a different problem for the same target audience, here are some questions to ask:

› Have you compiled the use cases for how your solution will be used to address this different problem?

› Have you trained your customer-facing teams (customer success, marketing, sales, services, solutions consulting, support, etc.) on the new use cases, how to speak to the problem, and how to resolve issues when customers use your solution in these new ways?

› Are new workflows required for these new use cases? Are they manual workflows, or will you automate them?

› Do you need new pricing for this new approach? Do you need new packaging? Do you need a new way to go to market?

› Does your solution need to change to support new go-to-market strategies, packaging, or pricing?

› Will this new approach require you to measure different metrics from your old approach? What would those new metrics be? How will you measure them? Who will report on them?

Targeting a New Audience

If you're pivoting to solve a problem for a new target audience, then you'll want to ask all the questions above, and also consider the following:

› Have you taken time to research your new target audience to better understand how they make decisions and how they like to work?

> ❯ Do you need different experts in your company experienced with this target audience?

> ❯ Do you need to change the language you use in your solution and customer-facing materials to better align with the language of this new target audience?

Selling into a New Country

If you're pivoting to sell into a new country, add the following to your list:

> ❯ Have you researched legal and regulatory matters in that country to see if you need to change how you operate?

> ❯ This might include changing how you collect data, hire and terminate employees, manage privacy, market and sell, package your products, offer support, and more.

Admittedly, all these questions take time to answer, and it may feel like thinking through all of these is just prolonging the chaos stage of the change you're trying to implement. That's a fair concern, because once you've started, the longer it will take you to complete a pivot, the more confused you'll make your employees, customers, and the market at large.

But skipping the all-in pivot and doing a half pivot is just going to waste more of your company's precious time and resources. As you're aligning everyone to pivot together, limit the chaos by communicating the work everyone is doing along the way. It may feel safer to stay quiet until you're well into your pivot, but all that does is leave people in the dark to come up with their own stories about how things are going.

Pivoting the Platform	Hitch's platform was built to compete against talent mobility features within larger human resources (HR) solutions. It had gotten too big and complex for the new skills intelligence pivot. To ensure the company would succeed in their new strategy, they needed to make the platform more modular, so customers could get value from a smaller initial investment, and then expand into more modules over time, following a "land and expand" sales motion.
Pivoting the Sales Strategy	Hitch had been selling to large global enterprises that had complex use cases and twelve- to eighteen-month sales cycles. Hitch couldn't afford to invest that amount of time to see if their new pivot strategy was going to work. Based on numerous customer conversations, it became clear there was a large enough market opportunity in the high-growth tech space with small US-based enterprises — which tended to be agile and willing to adopt new technologies quickly. Hitch realigned their sales team to target that market instead.
Pivoting Positioning	And of course, to capitalize on The Crack in the Market they'd identified, Hitch aggressively pivoted its marketing strategy and product positioning from talent mobility to skills intelligence, driving excitement internally and externally to spark a Transforming Idea.

Key Takeaways

---

1 A pivot is a change in nearly every part of your business — it's an all-in effort.

---

2 Talk to every department in your company to get their perspectives on how the proposed pivot will affect how they operate and what they need to do to support the change.

---

3 To limit chaos as you're aligning every part of your company to pivot together, communicate the work everyone is doing along the way. Don't leave people in the dark to come up with their own stories about how things are going.

Sidebar:
Friction < Value = Adoption

The concept of friction-to-value is important for product development and user testing strategies. It refers to the balance between the friction, or obstacles users encounter when trying to use a product or service, and the value they derive from it. It's a ratio that's crucial in determining whether users will adopt and continue to use a product or service.

It's also an important notion to consider when you're designing a pivot. If you don't gauge the friction-value ratio correctly, you won't get to business-market fit.

Here's how to think about it:

1. Friction represents any hurdles, difficulties, or barriers that users face when trying to use a product or service. Friction can come in various forms, such as a complicated or nonintuitive sign-up process, a steep learning curve, technical glitches, or slow performance. High friction can discourage users and make them abandon the product or service before realizing its value.

2. Value refers to the benefits and utility users receive from a product or service. It's what motivates users to engage with and continue using the product. The perceived value can vary depending on the user's needs and expectations. A product or service should provide enough value to justify the effort required to overcome any friction.

3. User Adoption is the process by which users start adopting a product or service and continue to do so over time. High user adoption indicates that a significant number of users have embraced the product, while low adoption suggests that users aren't finding enough value to justify the friction.

HIGH FRICTION / LOW VALUE	HIGH FRICTION/HIGH VALUE
LOW FRICTION / LOW VALUE	LOW FRICTION / HIGH VALUE

Pivot Wave Wrap Up

If you take anything away from this wave, remember a pivot is something that must be managed intentionally. A pivot is a stressful time. It's probably the first time your business really feels like it could fail. There's usually a way out if you have the perseverance, adaptability, and empathy to avoid the big mistakes that keep pivots from being successful.

First, be sensitive to indicators that it's time for you to consider a pivot, instead of waiting for one more customer, quarter, or feature to push you forward. Once you realize you need to pivot, don't start flailing. Generating idea after idea in hopes of one sticking is only going to drive frustration and panic among your team. Instead, be prepared to let go of the story you've been telling yourself about what your business is going to be. Revisit your assumptions, take a fresh look at what you thought were immovable truths, and assess what you got wrong, preferably with a trusted group of advisors. This helps you focus on where to look.

With that information, seek out market signals, listen to your customers, and challenge everything. Take inventory of all the good things you've built so far, such as your product, team, and customer and industry relationships, and consider how you can leverage these strengths for a new direction. Once you find your new vision and The Crack in the Market, it's time to go. Full speed ahead.

Bring your team along. Acknowledge the chaos they will inevitably feel and rally the troops around your Transforming Idea. And finally — go all-in. Inspect every corner of your company so everything points in the new direction.

The ability to successfully pivot is what separates promising startups from stagnant ones. If your pivot has been successful, you will have achieved repeatability and will be ready to face the Scale Wave.

The Third Wave

Scaling Rough Seas

The Scale Wave is a deep-water wave. It hits just as you're starting to feel like you've figured out how to captain your startup ship. You have customers who like and want your product, and you've figured out how to sell your product to those customers. Your sails are full, and you're sailing fast. And yet, somehow, parts of your organization are taking on water — quickly.

Pundits talk about scale in terms of numbers: point your compass at $50 million in annual revenue, or $100 million, or more; manage your company differently when you hit 150 employees, or 500, or more. But revenue and employee targets are very rough guideposts that don't help founders and leaders understand the real difference between Launch and Scale. Recognizing when the Scale Wave is about to hit you goes beyond metrics. Here's what it feels like.

> Your workforce is rowing in different directions. As a leader, you might not even realize how chaotic it feels for your teams because they're hiding it from you. Or you may not want to face the fact that it's chaotic. But it is.

> Teams are working at cross purposes, or even against each other. Employee satisfaction is starting to fall.

> Product velocity has slowed down, and new features are taking longer to release.

> The sales team is bringing in new customers, but it feels harder than it needs to be to keep your customer base happy. It's also harder to keep your (still relatively small) sales
team happy.

> Customer support tickets are racking up, and it's not clear that anyone has a plan for how to get through them all this year.

> Your board is asking you what you'll do to grow your revenue faster. Your team's only answer is to hire more people.

You keep hearing, "These are good problems to have." But the problems keep piling up.

As the Scale Wave hits you, you'll need to manage your products, your people, your customers, and your processes in a completely different way than how you have so far. In fact, the behaviors that made you successful as you conquered the Launch and Pivot Waves are the very behaviors that can sink you as you start to Scale.

> The behaviors that made you successful as you conquered the Launch and Pivot Waves are the very behaviors that can sink you as you start to Scale.

Don't make the mistake of believing that Scale is something you achieve in one move; it's a series of choices and adjustments as well as some seismic shifts. As long as your company keeps growing, you'll likely never fully emerge from the Scale Wave. You must create new structures and channels among your teams, with your customers, and within your market. These new scaffoldings will enable your business to expand its breadth and reach, rather than splintering into a bunch of small life rafts focused solely on their own survival. But as long as you're still on your journey, more waves will continue to hit you. You'll just be better at handling them.

A common mistake leaders make as they enter the Scale Wave is not being able to get out of the chaos that drives your teams to be in firefighting mode. **Mistake 10: Staying in a Constant State of Firefighting** is letting your company stay there.

In **Mistake 11: Managing Like a Startup Founder**, we focus on you. Because captaining your ship through the Scale Wave is mostly about leadership. And the biggest mistake you can make is continuing to manage like a founder as you crash into the Scale Wave.

Mistake 12: Staying within the Four Walls of Your Own Product looks at the mistakes you make when you fail to change your relationship with customers as you scale. You'll know if you're making this mistake if your customer satisfaction drops, your churn rate increases, and your sales team finds it harder to close new customers.

Mistake 13: You Treat Your Customers as Nothing More than Customers is continuing to treat your customers as nothing more than customers and the challenges of scaling when you do that. We describe one program you can employ to scale customer retention and growth without having to keep hiring more and more people — creating customer communities.

Finally, in **Mistake 14: Partnering as a Side Hustle**, we explore another program companies often attempt in order to scale their sales — partnerships. Like customer communities, well-navigated partnerships are a great way to scale your business without having to do everything yourself. But as the wave keeps crashing against your ship, many leaders are too preoccupied with firefighting to do anything more than treat partnerships as a side hustle.

You'll notice that none of our "mistakes" are about engaging in strategic brainstorming sessions or whiteboarding scenarios and future states. As operators who have done the hard work of navigating these waves, we've learned that strategy won't solve your execution problems. So read on for real execution challenges that can occur as you scale, and how to address them.

Strategy won't solve your execution problems.

We start this section with a story of a B2B SaaS company that hit the Scale Wave as they were looking to sell to larger, enterprise customers. Laura Marino was the Chief Product Officer at this company and walked us through what the wave felt like as it hit the product and engineering teams.

Story:
Rebuilding the Ship Before the Next Wave Hits

Laura Marino has been a product management executive at half a dozen technology companies as they've navigated the Scale Wave — what she calls their "teenage years." During the mid-to-late 2010s, Laura served as Chief Product Officer of a San Francisco-based SaaS company that had developed a software platform to help large organizations manage the complexities of relocating employees. The business had been operating for six years and had crossed double-digit annual revenues. Its founders were futurists, early to identify the future trend of employees becoming more mobile and technology allowing businesses to hire and move employees across the globe. Their relocation product was being used by a variety of customers, including a few large companies.

These large companies were asking for additional features to deal with the more complex workflow of their large operations. They also wanted the product to be flexible enough to allow integrations into existing parts of their business operations. In *Crossing the Chasm* speak, these companies were the early majority. They had specific needs they expected their vendors to meet, and existing workflows that new products had to fit within.

As Laura recalls, "Sales started getting ahead of the product" — selling features and functionality that were still in beta or, worse, not yet committed on the roadmap. (Ever seen that happen before? Perhaps in **Mistake 5: Scaling Too Soon**, during the Launch Wave.) That led to more bugs in the product, more support calls, and eventually, more unhappy customers. The company was hitting the beginning of the Scale Wave.

When Laura joined the company, the common refrain was the product and engineering team kept missing their deadlines.

There were several indications that the company was having difficulty navigating the Scale Wave.

> The technical teams were doing their best to keep up with new product requests from these new larger customers, but they couldn't keep up their development effort at the quality these established companies demanded.

> Every time a large new customer was brought on, engineering had to get involved to onboard the new customer. The business hadn't invested in the configurability and APIs needed to enable a services team to take on responsibility for onboarding new customers.

> Support calls started increasing. Many were routed to engineering to resolve, increasing the work on their plate even further.

> From handling fire drills from support to being dragged in to serve a particular customer, the product and engineering teams started missing their deadlines.

> Worse than that, the velocity of new features dropped — because building on top of a non-scalable platform takes a lot more time.

The business had hit a size that required the product to be re-architected for scale. The product hadn't "grown" with the business. That doesn't mean the product hadn't added more features and functionality — in fact, it had probably doubled its feature set over the previous years. But those new features were built on top of other features without considering what the growing monolith would be like to support and manage.

In a perfect world, your software product would be architected to support future expansion. But in the Launch Wave, you don't live in a perfect world. You live in a world of minimum viability and speed. The engineering team believed that a full re-architecting effort had to be undertaken immediately, so future development work could be sped up. The problem? Engineering estimated it would take about twelve months of re-architecting work, a period that could halt the company's growth in its tracks. (If you've worked at enough software companies trying to scale, you'll recognize "twelve months" as code for, "At this point, I don't know when it will be completed.")

Scaling is a fine balance of rebuilding your boat without reducing your speed so much that you capsize under the wave.

> Scaling is a fine balance of rebuilding your boat without reducing your speed so much that you capsize under the wave.

Laura's challenge was to chart a path to re-architect their solution for scale, while continuing to serve the growing needs of their expanding large customer base at the same time. She broke down her approach into seven steps:

1. Segment out specific problem statements
2. Map out dependencies
3. Prioritize engineering initiatives as part of the product roadmap
4. Deliver value every quarter
5. Recognize there will be unavoidable disruptions
6. Make room for testing and feedback
7. Beware the whale

Step 1: Segment out Specific Problem Statements

The teams started by breaking down the generic problem — it was taking too long to add more features and fix bugs — into more consumable segments. The results were specific problem statements that could be addressed separately.

› We need to build self-service configurability into the product, so engineering work isn't needed every time a new client is onboarded.

› We need to develop monitoring capabilities, so we can receive real-time usage metrics to help us identify bugs earlier.

› We need to focus first on rebuilding our user interface, written in an outdated programming language, because that will help us more quickly build out features that improve our user experience.

Step 2: Map Out Dependencies

Solutions to some problems often require other solutions to be implemented first. For example, the team couldn't scale or build out self-service configurability until they first re-architected their user interface. Instead of re-architecting the whole platform at once, the team mapped out the business needs to the parts of the platform that became dependencies in solving those needs.

Step 3: Prioritize Engineering Initiatives as Part of the Product Roadmap

The future can't be fixed all at once. The next step was to prioritize the changes that were most critical to the goals of the business, balancing them against changes that couldn't happen until others were put in place first. Laura pointed out:

"This is the core job of product management. Together, as a more united crew, we developed a four-quarter roadmap, balancing competing priorities, and refining it every quarter to accommodate learnings and changes in the competitive landscape."

Step 4: Deliver Value Every Quarter

Regularly delivering value meant that sometimes the team would prioritize a smaller project over a larger, higher-priority one to prove to the rest of the company they were re-establishing their ability to deliver value quickly.

Step 5: Recognize There Will Be Unavoidable Disruptions

With the problem statements broken down into more manageable segments, the dependencies to achieve them mapped out, the whole list prioritized, and the need to deliver value every quarter recognized, the team had all the elements they needed to create a product roadmap that served their scaling company. This effort should take about a month of intense work, not a quarter. Laura noted:

"No matter how well you prioritize and plan, there will always be unexpected bugs or critical customer requests that will need to be addressed. To account for that, we added a buffer to their delivery targets. However, these buffers weren't randomly generated. We leveraged historical data to estimate

the percentage of time we should expect in 'business as usual' efforts to keep the business running. This resulted in more reliable timelines and better expectation setting for customers and the sales team."

Step 6: Make Room for Testing and Feedback

You wouldn't — or shouldn't — take a newly built ship across the ocean on its maiden voyage without testing it first in safer waters. During the Launch Wave, you're selling to innovators and early adopters. They forgive bugs and missing features because they have a high tolerance for risk. Not so with the early majority. Laura and her team knew that they not only had to deliver value every quarter, but they also had to ensure higher quality to maintain and grow their larger customer accounts.

Step 7: Beware the Whale

Sales can't help but want to bring in the whale — that huge customer that could represent a deal that's three times larger than anything else you've brought in to date. When you're getting your product ready for the Scale Wave though, a whale can sink your ship. That large customer comes with outsized demands that can take over your product roadmap — that very roadmap that everyone just spent weeks (hopefully not months) assessing, negotiating, prioritizing, and scheduling.

When the whale came into Laura's company, product and engineering immediately became concerned about the amount of customer-specific functionality they were requesting. The work the customer wanted done wasn't aligned with the overall direction of the product, and based on the teams' analysis, it was going to require more than 15 percent of the total engineering bandwidth to complete. Armed with this data, Laura was able to convince the rest of the executive team to pass on the opportunity.

The hardest thing a scaling company has to learn is when and how to say "no" to a large customer opportunity.

Mistake 10: Staying in a Constant State of Firefighting

You've now crushed your MVP, PMF, TAM, SAM, SOM, and all the other TLAs (three-letter acronyms) you can remember. It's time for you to rejoice in your relentless pursuit of growth. But as your startup scales, your teams face more and more pressure to meet the dynamic internal and external demands of your growing business.

Laura's story is representative of the experience most scaling companies go through. With more features come more bugs to fix. With more systems come more workflows to fix and integrations to support. With more customers come more customer requests and support tickets. With more employees come more policies and employee relations issues. With more revenue comes more accounting. And with more scale comes more tech and operational debt to fix, more deadlines missed, and more crisis moments.

Darrell (UserTesting's co-founder) recalled the start of their scaling stage. UserTesting had built a sales team led by their third co-founder, Chris Hicken. They were heading into double-digit annual revenue territory. "We were getting more support tickets now that we had more customers," Darrell remembered. "There were a thousand problems that had to be solved."

If you aren't prepared when you hit the Scale Wave, you start looking more like firefighters than air traffic controllers. Firefighters solve problems after they arise. Air traffic controllers set flight paths in advance, anticipating needs and constraints, and then constantly monitor operations and communicate with flight crews to ensure everyone reaches their destination safely — no firefighting needed.

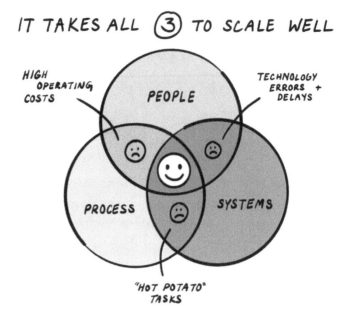

IT TAKES ALL ③ TO SCALE WELL

HIGH OPERATING COSTS

PEOPLE

TECHNOLOGY ERRORS + DELAYS

PROCESS

SYSTEMS

"HOT POTATO" TASKS

Stop Being the Emergency Response Team

As your startup grows, more problems are generated than there is time available to deal with them. This is the firefighting stage, and every company goes through it. Sadly, some never get out of it. That's because some people love being firefighters. Firefighters are the ones who swoop in at the last minute when someone identifies a crisis, working long hours to resolve it. And then the company showers them with accolades for their heroics — making each crisis the fault of management as much as the firefighting employee.

In a blog posted by Milliken, an industrial manufacturing company, they assessed that under traditional management "employees spend 60% of their time solving daily unforeseen problems that we call 'firefighting.'"[1] Business leaders have developed an unhealthy habit of rewarding reactive firefighting instead of the less glamorous, but more proactive, effort of thoughtful advanced planning and scaling.

---

1 Gaspar, "Operational Excellence," Milliken.

Scaling means spending less time in reaction mode and more in pro-action mode. If key people on your team love being firefighters just a little too much, you'll need to redirect their energies or let them go so as not to weigh you down as your ship battles through the Scale Wave.

Scaling means spending less time in reaction mode and more in pro-action mode.

Stop Saying NO and Start Actually Doing NO

As you're scaling, your product team is going to have to fit in a number of efforts they haven't needed to invest in before. If you're a software company, your product scaling efforts will include new projects. They'll have to meet new compliance standards, implement enterprise-level authentication protocols, uplevel their security posture, build out APIs, perhaps internationalize the product for use in other languages, and as Laura's team had to do, re-architect the product, so it can scale to support the next wave of up to ten times the number of customers you currently have. And do all this while continuing to support requests from customers who want more features and functionality in your product.

Additional scaling projects will be needed from your sales team, your marketing team, your customer success and support team, and your back-end operations and administration teams.

That's a lot of new work. And unless you've been paying a large group of high-performing employees to just sit around until you're ready to scale, you'll have to learn how to prioritize.

Prioritizing could be one of the most overused and under-actioned words in business. We've been in a lot of meetings where we said we were going to prioritize some project or another. What that inevitably turned into was accelerating the deadline for the "priority" project while still working on all the other projects in the background. Instead of "prioritizing," scaling companies need to start "discontinuing." Like "end-of-life"-ing a product, discontinuing a project should have an entire project plan and communication plan associated with it.

 Discontinuing a project should have an entire project plan and communication plan associated with it. You must do stuff to end stuff.

You must do stuff to end stuff. Companies that can't make the hard decisions to stop doing less critical activities while they're learning how to implement new scaling initiatives will end up drowning under the gush of breaking processes that can't support the increased operations of the business.

Don't Be a Grasshopper

Grasshoppers shed their skin half a dozen times within a six- to nine-month window during their "teenage" years. We've seen founders trying to grow who quickly shed their last strategy and come up with a new idea five or six times during a two- to three-quarter period. Frequent change is good for grasshoppers — and for startups during the Launch Wave in the middle of experimentation. But it's bad for scaling companies. It's impossible to execute against a moving target. And constant change creates more firefighting. Scaling means being able to focus on executing a few goals long enough to see if the strategy works.

How long is long enough? It's a lot longer in the Scale Wave than it was in the Launch Wave. However, nothing should take longer than a year. As a scale leader, you're threading the needle between being a grasshopper and being, let's say, a Greenland shark (which grows about a half-inch per year as it scales to an age of over 250 years). How you thread that needle will depend on the size and complexity of each project you must undertake.

Small and Simple Projects *(Such as Updating Your Policies and Procedures)*

Just get this done. Andy MacMillan, the CEO who Darrell brought on to scale UserTesting, would say, "Don't give it to a blue-ribbon committee to plan out." This doesn't need a Gantt chart — it needs one or two people to own the project, clear their schedules, and get it done in thirty days or less. These are projects that've been done before by many companies, aren't proprietary in nature, and are usually good enough if they get to about 80 percent perfection.

Larger and More Complicated Projects *(Such as Upgrading Your CRM System)*

Hire someone who has done it twice and did it well the second time. This person already knows the twelve-step process, so you don't need to reinvent it yourself. Consultants generally won't work — they aren't there to influence all the business process changes that need to occur as part of the overall project effort.

Be realistic about the deadline but have a sense of urgency — there will never be a future moment when it gets easier to make the change. Give this project one to three quarters to complete, depending on the extent of the operational debt you incurred before starting the project. But make sure your been-there-done-that hire shows early wins within a quarter.

Huge and Complex Projects *(Such as Re-architecting Your Product for Ten Times Scale)*

These projects are often specific to your business and how it operates. If you hire someone new to run it, they're going to take some time to ramp up before they can really begin to sink their teeth into the project. And that's assuming you hire the right person. How many times have we assumed we were just one perfect hire away from solving all our problems? You need to identify a project leader with a great combination of strategic outlook and operational project management. If you don't have that person, you're rolling the dice with a new hire. In either case, the mistake many companies with little scaling expertise make with huge projects is trying to plan out the entire project before beginning to tackle it.

> How many times have we assumed we were just one perfect hire away from solving all our problems?

Like in your Launch Wave, huge scaling projects require small milestones and agile course correcting, all while maintaining a constant focus on your North Star. But unlike your Launch Wave, to scale well you'll need to invest in some long-term efforts that might not show a return within a few quarters. Underestimating the effort these kinds of projects involve — a common mistake — creates all kinds of problems from staffing to funding to poor planning. As a scale leader, your job is to understand when you're committing your company to a huge and complex project, get the right leader for it, and avoid changing direction mid-course. Resist your grasshopper tendencies.

Preparing to Fight Fewer Fires

During the Launch Wave, you absolutely should not build scalable processes. Riding the Launch Wave is about trial and error, then speed to get to the next trial and learn the next error you made. Now, as you turn your ship into the Scale Wave, you must strategically pick areas of your company and start doing the opposite — replace your hacked together solutions with more robust, reliable systems.

Newsela was an education software and content startup that saw a deep pain point schools would have following the United States' implementation of the Common Core educational initiative, which launched in 2010. Common Core required teachers to use more nonfiction materials — such as essays, journalism, opinion pieces, and speeches — in English Language Arts classes. Teachers weren't given the support they needed to access those materials efficiently, and Newsela stepped in with a software-automated solution to deliver content that met this regulatory need.

In the early days of the company, two of Newsela's co-founders, Jennifer Coogan and Dan Cogan-Drew, would spend hours each day reviewing a list of all the new teacher registrations they'd received that day. No matter how many they received, they'd look at each one, and if they saw more than one registration from the same school, then they'd send out a personal email to the teachers from that school who had registered, thanking them for signing up and offering their help. These emails would get the teachers talking together about what they liked about the service, and then other teachers in the school would hear about it.

These personalized, small group emails, along with other tactics the founding team employed, were a brilliant Launch Wave marketing hack for getting teachers to spread the news about this fresh solution and drive more sign-ups. But as the company scaled, these hacks that served them so well were turning into fire-fighting efforts.

Newsela couldn't scale by continuing to only sell on a school-by-school basis. This go-to-market approach was an expensive way to get small customers (schools) that easily churned because they had to be onboarded one at a time.

Like every startup, there was no shortage of areas where Newsela felt it could invest in improving its operations. But startups can't address multiple huge, complex problems at the same time. Newsela chose their go-to-market motion as the area they needed to scale next. And to prepare for this change, they hired their first professional sales team focused on selling at a school district — and eventually state — level. They proactively let go of their old sales approach that had worked so well thus far. They created new measures for success, including reducing their churn rate, which had been high when they were selling at the school level.

As you invest in scaling your business, how do you know if the investment is paying off? You'll know you're getting more proactive and less reactive when you see yourself rewarding the right behavior and measuring the right metrics.

You Learn to Recognize and Reward Proactive Efforts

In Laura's story, the advanced analysis, planning, and implementation of a product roadmap that delivered value every quarter didn't come with Hail Mary passes and diving catches. It came with hard work and strong execution — practices that can easily go unnoticed (and often are).

When you learn how to recognize — and start to reward — proactive, strong execution over firefighting, your employees will change their behavior to align with that recognition.

Your Metrics Show You're Getting the Right Stuff Done

It's beyond the scope of this book to cover how to implement performance and project management metrics in your organization. For the definitive perspective on project management metrics, look at books by Harold Kerzner (we recommend *Project Management: A Systems Approach to Planning, Scheduling, and Controlling*, and *Innovation Project Management: Methods, Case Studies, and Tools for Managing Innovation Projects*). For our purposes, we suggest implementing metrics in three key areas:

1. Product performance: Most companies measure product team performance metrics — how many bugs are fixed, how many features are implemented, how many engineering hours are invested. Metrics like these encourage firefighting activities and don't separate priority work from non-priority work. Instead, measure how your product impacts your business goals. For an example, see the **Sidebar: Product Performance Reviews** at the end of this chapter.

2. Project performance: Through the Scale Wave, you'll be introducing new projects that improve your ability to scale (some small and some very large and complex). Companies drown under the Scale Wave because they fail to track project performance. Projects go off track, leaders come up with bright new shiny projects to pursue, and employees regress into emergency response teams to clean up.

3. People performance: In the end, businesses are the sum of the effectiveness of their people. There are too many articles, blog posts, and books written about employee performance management, focusing on career development, compensation management, performance reviews, training, and other HR solutions. These are all tools, but if you don't point your tools in the right direction, then you still aren't going to know if you're successfully moving out of firefighting mode. Up next, **Mistake 11: Managing Like a Startup Founder** is about avoiding the mistakes that point your people in the wrong direction.

Key Takeaways

1. Firefighters solve problems after they arise. Air traffic controllers set flight paths in advance, anticipating needs and constraints, and then constantly monitor operations and communicate with flight crews to ensure everyone reaches their destinations safely. To scale, you must help your team become air traffic controllers instead of firefighters.

2. Prioritization exercises often turn into accelerating deadlines for the "priority" project while still working on all the other projects in the background. To scale, you must cancel some projects before you prioritize others.

3. It's impossible to execute against a moving target. And constant change creates more firefighting. Scaling means being able to focus on executing a few goals long enough to see if the strategy works.

4. Scaling means learning how to recognize and reward the right behavior and measure the right metrics.

Sidebar:
Product Performance Reviews

Product development is one of the harder areas of the company to assign meaningful metrics. Many software engineering departments measure their "performance" by assigning some engineering velocity metrics in project tracking software. These metrics are often fungible.

Early-stage scaling software companies often spend up to a quarter of their operating expenses on their product and engineering teams. That's a significant investment and warrants a serious commitment to more instructive performance indicators.

All performance indicators should be developed considering the goals of the business. For example, if the business is looking to grow through the addition of new customers, then teams — including product and engineering — should gear their efforts toward attracting new customers. If, on the other hand, the business is looking to retain and expand its business with existing customers, that could warrant different efforts. (And, of course, if the goal of the business is to do everything all at once, then it doesn't really matter where efforts are directed, because if you try to do everything, you'll end up accomplishing nothing.)

How can you map the efforts of your product team against your business goals? Let's say your company's current goal is to retain more users (reducing churn). You can invest in customer support to try to reduce your churn rate. But support tickets are a lagging indicator of possible churn. So instead, your product team decides to launch some new features intended to help customers better use the product so that churn decreases.

For example, they could launch the following three features:

1. An analytics dashboard, so customers don't have to download data to analyze it
2. An integration with another piece of software in the user's tech stack
3. A few in-app prompts that appear the first five times the user opens the app, suggesting custom use cases based on that user's profile

After a year on the market, the company's customer churn rate decreases. The company is still not at its churn rate goal, but it's closer. You celebrate. In the second year, the product team adds a number of additional analytics to the dashboard (choosing these features over others in the backlog) as well as more integrations. Another year passes, but the churn rate doesn't change. You scratch your head.

What drove the initial churn rate reduction? Without having instrumented the product to see which features were being used, your product team can't answer that question. Perhaps the dashboard they built was barely used, and the integration was too complicated to set up. It could have been the in-app prompts that were helpful in getting users to continue using the product. But without smart product metrics that can drive accurate product performance reviews, you'll never know.

Mistake 11: Managing Like a Startup Founder

Continually fighting fires is one mistake scaling companies often make. Failing to change how you manage your own time and leadership capital as the Scale Wave swells is another. In this chapter, we'll talk about your evolution from founder-leader to executive leader. This can impact almost every aspect of your role. It requires you to reframe the ways you motivate your team, focus your priorities, hire appropriately for this stage of growth, train your managers, and build safeguards to preemptively avoid points of failure you might not have seen coming. As the company roster grows, you'll have to work proactively to prevent conflicts between your old guard and your new hires, so the whole crew is sailing in sync. In short, you'll need to rewrite your own job description even as you're steering the ship.

In the Launch Wave, you became good at motivating your team. As a strong leader, you made sure you knew everyone personally — something you could do with a team of fifty-ish people or fewer. Because you could talk to each of them individually, you made sure they were all rowing in the right direction.

Sadly, the skills you used to motivate your team of fifty don't translate to teams of 150. Bob Tinker, CEO of MobileIron, recognized this as he learned that companies tend to break at 50, 150, and 450 people. Even the best, most energetic leaders reach the point at which they can't keep up with all the new hires the company is bringing on. And if you're keeping up with that, then there are other things you should be doing that you're neglecting.

To conquer this third wave, you must stop managing like a founder-entrepreneur — even though we just spent dozens of pages telling you that's exactly what you have to be. Because continuing as a startup founder during a Scale Wave will have you facing five common slipups. In the rest of this chapter, we cover these five blunders in more detail.

1. You'll continue to try and motivate each employee in a personalized way, but you won't be able to keep up with your growing employee base.

2. You'll continue to experiment and come up with new ideas. And you'll exhaust your employees in the process.

3. You'll hire wrong. You may continue to hire generalists. Or you may start hiring what we call "Career Officers." Neither will work. But when you do learn how to hire, you'll be faced with territorial infighting between your original teams and your new teams.

4. Or maybe you won't hire. And you'll have SPOFs (single points of failure).

5. You'll have hard decisions to make and so many more people wanting to provide their opinions on how to make them. You'll be stuck between satisfying everyone's need to be heard, controlling everything like you used to be able to, and being overwhelmed with so many changing relationships.

Motivating @Scale

Motivating a scaling organization is a different challenge than motivating a small, close-knit team. To motivate at scale, you, as a leader, must learn to communicate at scale. What does communicating at scale look like?

› You have a 10X story to tell, not a 10 percent target to discuss. As Astro Teller, CEO of X (Alphabet's "moonshot factory") said, "It's often *easier* to make something 10 times better than it is to make it 10 percent better. . . [Because] when you aim for a 10X gain, you lean instead on bravery and creativity,"[2] not just grit and perseverance. Communicating 10X stories inspires and gives purpose to your team.

> **You tell your story repeatedly at company meetings.** We always remember our own stories and strategies. But general wisdom suggests people need to hear a message seven times to remember it. Brad D. Smith, former CEO of Intuit, often said, "Repetition doesn't ruin the prayer."

> **You tell the story to the broader industry.** When you were in the Launch Wave, we told you not to spend money on PR. Now it's time to spend. But do it right. Be consistent in your story, both inside the company and outside of it. When your employees see the same message they hear internally being discussed in industry publications, they get more excited about being at your company.

> **Communicate more than just your 10X story internally.** Employees want to feel like they're trusted with information about the business. Communicate your forecasts, plans, and how the business is doing against them. Communicate changes in the organization. Communicate when the going gets tough as well as when you achieve success.

Employees want to feel like they're trusted with information about the business.

Creating a sense of team camaraderie and belonging at scale requires intentionality. It's a skill set that's very different from the skills you've been exercising through the Launch Wave. You've been solidly, almost rigidly, focused on imagining the future of the business with a few trusted peers — so much so that it's easy to forget to build a regular cadence of bringing along the employees

---

2 Teller, "Google X Head on Moonshots."

who are supposed to make that future happen. Here are four ways that can go wrong.

1. A CEO of a 300-person company doesn't remember to send out communications to the entire organization when they hire a new executive (or fire an existing one).

2. A senior VP managing a 100-person team gives one strategy presentation to the team. Six months later, the team is complaining they don't understand the strategy they're supposed to execute against.

3. A CEO of a 400-person company that's starting to get external visibility is getting interviewed at conferences, on podcasts, and in publications. Employees come across some of these interviews through their own newsfeeds. They're never covered internally.

4. A CEO of a company that just prevailed through a Pivot Wave realizes the journey is going to get even harder as the Scale Wave approaches but doesn't feel comfortable addressing it with their employees. So they don't.

Henry Schuck, the founder and CEO of ZoomInfo, met this opportunity in an email he sent to his team shortly after the business went public in June 2020, at the height of the COVID-19 pandemic. He described his messaging during a podcast episode of *Invest Like the Best*, hosted by Patrick O'Shaughnessy.

"The gist of the message was we're building a championship team here at ZoomInfo. And not everybody is cut out for a championship team, and I use the analogy of the Pittsburgh Pirates and compare them to the Boston Red Sox . . .
"And every year, [The Pittsburgh Pirates] show up and they play the game, and they're never in the hunt for a championship title . . . It's a nice place to work. They make pretty good money . . . And then there are

other teams like the Boston Red Sox or the Yankees who every year, year in and year out, are focused on building a championship team, and they're always on the hunt for a title . . .

"There are lots of Pittsburgh Pirates out there that you could go and have a really comfortable existence at and not feel sort of this everyday drive to be the best to win a championship. And Pittsburgh Pirates are good people, and you can go work at that company, and you won't feel the pressure of what we're trying to build at ZoomInfo. But if you come in every day and you think that you're going to have the Pittsburgh Pirates and win a championship, it never happens . . . People felt really great about the IPO and the outcome of the IPO. But it had to be really clear that the IPO was just one step in a much bigger journey . . .

"If we got so caught up in how great we were when we IPO-ed just one little moment in time, [then] we forget that there's a much longer road in front of us where we have to continue to perform and continue to win championships. And look, I think there are places for lots of different types of professionals in business, but you should align yourself to the kind of team that you want to be on. Do you want to win a championship? And do you want the pressure and the performance management that comes along with winning a championship, because that's what you're going to get here, and it may not be what you get somewhere else."[3]

Henry made a purposeful decision in how he chose to motivate his team. And by doing that, he realized that some of his employees would choose not to stick around. But for the people who would stay, this was the 10X message they wanted to hear. And that's the team Henry wanted to lead.

---

3 O'Shaughnessy, "Henry Schuck."

Focusing @Scale

At this point, it's also time to throw out the idea machine that keeps generating new ideas and focus on the things that work. You've succeeded in launching and pivoting because you saw the future and became excellent at testing every new idea you had to meet that future need. Well, the future you envisioned is now. Now you must take the ideas that have proven successful and execute them — at scale. If you keep generating new ideas all the time, you run the risk of overheating your already-stretched organization. You'll suffer from continuing to pursue too many initiatives, leaving your teams unfocused and unable to deliver on any successfully.

That doesn't mean you shouldn't take on any more new projects. As you'll see in **Mistake 13: You Treat Your Customers as Nothing More than Customers** and **Mistake 14: Partnering as a Side Hustle,** building partnerships and communities are huge new projects that are critical to continuing your scale journey.

As we explained in **Mistake 10: Staying in a Constant State of Firefighting,** with more features come more bugs to fix. New systems bring more workflows to fix and integrations to support. More customers mean more requests and support tickets. More employees mean more policies and workplace issues to sort out. And more revenue brings more accounting. So, it's critical you allocate the vast majority of your resources to scaling what you've already proven to be a business opportunity. Like many things in life, we can apply the "80/20" rule to scaling versus innovating. When you're in the Scale Wave: spend roughly 80 percent of your resources on your scaling activities (running the current business more efficiently) and 20 percent of your resources on new initiatives (a second product? growing an ecosystem? building a successful customer community?). The 80/20 rule is a helpful guidepost. There are times when your business will need to spend 100 percent of your time scaling your current business. And as you tame the Scale Wave, you'll free up more resources for new initiatives, having to spend less of your company's time on improving processes, systems, and teamwork.

Great scale leaders do four things well with their teams to help them focus at scale.

1. They help their teams allocate roughly 20 percent of their time to supporting new opportunities, leaving 80 percent of their time to running the business.

2. They ensure that these 20 percent of resources are focused on supporting the same set of opportunities, rather than each team pursuing their own separate opportunities that help change the business.

3. They understand that 80 percent of the company's resources (in most cases) are needed to run the business, which means they limit change, prioritize just a few new strategic opportunities at a time, and complete or abandon one strategic opportunity before adding a new one for the business to implement.

4. They align to identify goals for every new opportunity they undertake and ways to measure progress against those goals. They also report on achievement against those goals on a regular basis.

Great scale leaders limit change, prioritize just a few new strategic opportunities at a time, and complete or abandon one strategic opportunity before adding a new one for the business to implement.

Hiring @Scale

Generalists vs. Experts

You're not the only one who has to change right now. As you navigate the Scale Wave, your employees will be shifting from the generalists who have had to do a little bit of everything to the specialists who are able to create high-performing expert teams in a particular functional area.

This is tricky, because some people will feel like their roles are being diminished as they move from having a hand in everything to a more specifically defined role. People who are used to being in every meeting might feel shut out when they're only invited to the meetings specific to their function. As Molly Graham describes in First Round Capital's blog, *The Review*, about her commandments for scaling startups, "If you personally want to grow as fast as your company, you have to give away your job every couple months."[4] As a leader, it's your challenge to help the generalists see the value they can provide in becoming specialists. If they can't make that transition, you should help them move on.

At the executive level, the calculus is a little different. You need people who can be both generalists and experts at the same time. Here's why. The best executives are the ones who are great at some functional area. They've got the expertise, and they know it deeply. And they've had a sufficient breadth of experience in the industry to know enough about all the other areas of the business. If you're great in one area, but you haven't really spent time getting to know how the rest of the business works, you may be a great functional leader — a senior director or VP — but not yet an executive.

 The best executives can go deep in a functional area while also having sufficient breadth of experience across most other areas of the business.

Real executives help captain the whole ship rather than worrying about their own little life raft. We know this, because

4 Graham, "Give Away," interview, *The Review* (blog).

when we were earlier in our careers, we were only focused on one of those little life rafts.

Mona recalls a time many years ago in her first executive role. She had the title of General Counsel in an early-stage startup trying to scale during the 2008 financial crisis. The economic crisis started in the US but spread to many countries across the globe. Bank failures drove down investor confidence, which destroyed credit availability, which restricted spending, and then caused businesses to collapse. Like many other businesses, Mona's company wasn't hitting its revenue goals.

Whenever there's stress in an organization, whether internally or externally imposed, teams have a habit of playing the blame game. And this situation was no different. Mona and the Chief Revenue Officer started butting heads. The CRO said the legal department was stopping deals from closing. And Mona blamed the sales team for not executing a competent sales strategy.

They were at it for a little while, until the CEO stepped in. The CEO was a decade younger than both of them, and this was his first management role. Yet impressively, he acted more like an executive than either Mona or the CRO. He pulled the two together and said, "You know, I didn't hire you to be functional leaders. I hired you to be executives of this company. And that's what I expect you to be. So, get together and figure out how to make this company succeed. And stop blaming each other."

The message was clear: "I don't know whose problem it is, and honestly, I don't care. But I'll fire you both if you don't figure it out." The CEO was right. They weren't acting like executives. They were acting like functional leaders.

The two of them got together over a few beers, talked it out, and started working together to meet the company's goals of both sales and risk management. And Mona has never forgotten the wise advice she received from her first CEO.

There are a lot of great functional experts out there. Imagine, for example, a technical genius CFO. But all they care about is the company's financial statements. They're going to prepare your forecast. And they're going to get that audit done. And, dammit, you're going to hit every industry financial ratio that applies to your business. But is your company ready to hit those ratios? Is that the best thing for the business at this point in its journey? Who knows? That's not their problem. And that's not an executive.

And then you've got the other extreme — the person who knows a little bit about everything but can't go deep on anything. They can prepare a slide deck that presents a great long-term strategy, or perhaps even an Amazon six-pager. They can provide templates for people to complete and ask for metrics to help them prepare. And in a large enough company, such a person can probably succeed if they hire a strong, functional VP to manage each team that reports to them. That adds another layer of management — an expensive one. But if that VP is weak, and the team isn't performing, your generalist executive won't be able to identify the problem, or perhaps even sense that there is a problem, until things are very bad. They won't be operational enough to know when the ship is starting to take on water. Placed in an executive role, you'll end up with more of a consultant — who likely can't own getting complex projects to completion.

Jeff Bezos's Amazon six-pager memo format originated in 2004 and is highly regarded as an effective communication standard for presenting ideas and running meetings. Rather than using slides, meeting participants are given a printed copy of a six-page memo that introduces the topic of the meeting. They have twenty to twenty-five minutes at the beginning of the meeting to read the memo and note down any questions or feedback on the printout; no one reads the document before the meeting. After discussion, participants hand in their copies, and the meeting leader updates and recirculates the document as a final version.[5]

5 Poschenrieder, "What is an Amazon," *Six Pager Memo* (blog).

As you scale your team for the next wave, you want to build an executive team in which each person has the depth — not across every area but at least across one functional area — and also the breadth of experience that drives them to fruitfully engage across the entirety of your increasingly complex operations.

Career Officers

For companies trying to scale, it can be tempting to make expensive leadership hires from big, established companies with impressive brand names. These hires come with impressive titles, usually with the words "global" or "worldwide" in them. They've managed large teams and even larger budgets. Surely, helping your company scale will be like an easy summer sail for them. And this may be true — sometimes.

But what you have to watch out for is Career Officers — leaders who are used to working in the Corporate Industrial Complex. They're great in that environment, where the only way to manage at such a huge scale is through process, where critical skill sets include navigating bureaucracy, and where when you need something done, there's someone, somewhere in the Complex that was hired to do it (and nothing else but it). It's been so long since a Career Officer implemented something themselves (other than a slide deck or a presentation) that their automatic reaction is to find someone else to get it done. They give out work — and make sure it gets done — but they don't do it themselves.

Once you've conquered the Scale Wave, by all means, hire the Career Officer. But before that, don't. Career Officers are trained to deliver in years. But in the early stages of the Scale Wave, your business still needs to be nimble enough to maneuver through the wave in months, or even weeks.

Career Officers are trained to deliver in years.
But in the early stages of the Scale Wave,
your business still needs to be nimble enough
to maneuver through the wave in months, or
even weeks.

As you try to move from creating a group of individual heroes in your company to commanding a series of heroic teams, it can be hard to find your balance. Some teams will still be operating like the Wild West. Others might build mature, process-oriented operations. Neither one might be perfect for where you are in your journey. While having either one exclusively isn't ideal, having both kinds of teams trying to row to different drumbeats at the same time is the fastest way to create conflict and sink your ship.

In addition to hiring the right combination of generalists and specialists to your leadership team, you also need to hire the right balance of process creators and rule breakers. If you're at this stage in your company's journey, the following story might hit close to home for you.

In addition to hiring the right combination of
generalists and specialists to your leadership
team, you also need to hire the right balance
of process creators and rule breakers.

We once advised a mid-stage technology company that was struggling to scale. It had gone through a year in which they'd grown their employee base quickly without putting enough constraints and processes in place. An executive managed to hire their friend at a huge salary for an ambiguous role without having to clear it with anyone. The CEO knew their internal execution was out of control.

So the CEO hired a few senior executives from some very large companies to put processes in place to help them scale. Another year went by. Now they couldn't get anyone hired because the job requisition approval process was so long and complicated that hiring managers couldn't get through it. Instead, they just worked harder to cover the deficit, and made their teams work harder, too. Similarly, vendor approval became a complicated multi-step process that had the unintended consequence of discouraging employees from onboarding vendors to implement needed systems. The company couldn't hire the people or implement the systems they needed fast enough. The ship was taking on a scary amount of water.

Our advice to the CEO was that they had over-indexed and hired Career Officers. The CEO eventually made the hard decision to implement a leadership change. The new leaders each had a combination of large company and startup experience. They knew how to establish industry-standard processes — they didn't need to make it up as they went. They also understood when it was necessary to set aside the process and act with the urgency and agility the business required.

It was still tough going for yet another year because the leadership team now had a combination of process worshipers, free-wheeling eccentrics, and methodical moderates. Rebuilding a functioning workforce that could meet the Scale Wave was a necessary investment, but having to do so sucked up a lot of effort from growing the business.

SPOFs @Scale

One of Mona's favorite moments was reading a report from a consultant her company once hired to help implement a complex ERP (enterprise resource planning) system. The consultant's list of project risks included "the concentrated combination of responsibilities owned by a few employees and the uncommon organizational chart." That was a (sort of) nice way of saying that the business had a heavy reliance on a few key people because they were the only ones who had knowledge critical to the project, and sometimes kept it closely guarded. These few people are what we call single points of failure — SPOFs.

> Single points of failure (SPOFs) are key people a company has heavy reliance on because they're the only ones with knowledge critical to the business, which they sometimes keep closely guarded.

To avoid drowning under the Scale Wave, it's no longer okay for one person in your organization to be the only one capable of handling a critical function. If there's only one member of your crew who knows how to provision customers, for example, you're going to have a lot of unhappy customers when that person falls ill or gets stranded at port. People naturally want to own their unique roles, and some might feel like they need to protect their turf. Making sure they learn to share, by documenting their historical knowledge and bringing someone else on to play back-up, is critical.

When your company is small, everyone on board is a potential SPOF. It's not just that one person knows how to do something that nobody else does. Everyone knows how to do about five different things no one else knows how to do.

At some point, you start hiring more people because you're getting some traction. But you're finding it hard to get your new engineers to be productive because all the intrinsic knowledge about your millions of lines of code is wedged in the head of your first star developer. It's not commented, or documented, or broken out into modules. It's incomprehensible to anyone but your rockstar coder. You've come face to face with your SPOF.

SPOFs don't just exist in your product organization. We've spotted them everywhere. Maybe you have five people on your sales team, and they're each responsible for your twenty largest customers. They're the only ones talking to the customers, and if you lost one of them, you would lose all that contact with the customer.

Possibly you solve your SPOF sales problem — you now have forty sales reps and each prospect is managed by a team of two or three people. But once a new account has closed, you've got one customer success person responsible for that account. What if they leave the company? Suddenly, your customer's emails to your company start to bounce. True story, unfortunately.

When you get to two hundred people, you're not going to be able to ferret out SPOFs on your own. Your teams need to be the ones that identify where these SPOFs are hiding. Your job as the leader is to help your teams feel comfortable identifying single points of failure. Most importantly, when SPOFs are identified, they should be recognized for their work and brought into the process of helping to train others on how to do that work. Your SPOF, the one who is responsible for this critical function that nobody else knows how to do, has to feel like they're part of finding the solution. Otherwise, they're going to become protective of their job, and just like that, they'll clam up for fear of losing their leverage.

> Your job as the leader is to help your teams feel comfortable identifying single points of failure.

Conflicts @Scale

While all of this is happening, you'll be hiring like crazy if you're scaling well. And as you do, you'll inevitably end up with an "Old Guard" of early employees and a "New Guard" of newer hires. It's important to help them integrate well with each other, or else you'll end up with two different cultures, likely at odds with each other.

You'll find yourself starting to invite more senior members of the New Guard to meetings that the Old Guard used to attend. The longer-term employees will inevitably feel like they're losing their seats at the table. The Old Guard got used to contributing their ideas and insights on all aspects of the business. They became used to having a say in many areas. Your New Guard will likely be more experienced in their respective functions (otherwise, why are you hiring them?). These new leaders will start claiming bandwidth at team meetings, making your long-term employees feel overshadowed and squeezed out. And truth be told, some of them may no longer have a place in those conversations.

You can't just keep adding more chairs to the room, or your meetings will start achieving the productivity of many governments. Set expectations early that these seats aren't tenured positions.

Still, it's important to help the Old Guard continue to feel engaged and valued, even if they're no longer included in the Monday morning leadership meetings they'd grown to love (or perhaps hate, but nevertheless feel entitled to attend). To do this, invest much more than you think is appropriate in ongoing communication. Share updates regularly and create opportunities for larger teams to ask questions about the business. Be clear: these aren't opportunities to change decisions — they're opportunities to understand the company's strategies and plans so everyone can be aligned. Invest time in helping them understand the context of scaling and that it doesn't mean they're less important, even though their roles may feel narrower than they used to be.

Some of your Old Guard will adapt. Others won't. Some of the choppiest waters you'll hit during the Scale Wave will result from your failure to deal with those who can't adapt to this change.

> Some of your Old Guard will adapt. Others won't. Some of the choppiest waters you'll hit during the Scale Wave will result from your failure to deal with those who cannot adapt to this change.

As you move from managing like a founder to stewarding your company through the Scale Wave, you and your managers will need to keep a close eye out for employees who aren't able to adapt to your changing business needs. Because team dynamics are multiplicative, not additive. (Math lesson! With addition, 5+0=5; with multiplication, 5x0=0.) One underperformer on a team doesn't just mean that one person's job isn't getting done while the other five team members are thriving. It means that the entire team's work, along with its motivation and sense of engagement, is likely being dragged down by that one person who isn't scaling with you.

Your New Job @Scale

We don't often talk about the founder CEO's relationship with their executive team. There's a different tension in that relationship than there is with non-founder CEOs. This is especially true if there are also other founders on the executive team.

If the founder CEO is the company's technical genius, other members of the executive team might be walking on eggshells around them. Perhaps the product needs some fixing or updating at this point. But the rest of the senior team might be hesitant to offer the kind of honest critique the situation requires. The founder birthed this technology, and it's hard to call someone's baby ugly — especially when that person is your boss. People at the top might hold back on saying what needs to be said. We've all been there.

The founder CEO's role is complicated in other ways as well. If they were successful in closing the company's first series of significant deals, then they may have come to consider themselves the sales expert in the company. However, that doesn't mean they should be the one continuing to close deals as the company crawls toward double digits in millions of dollars of annual revenue. It's time to bring in a brilliant sales executive, and then, just as importantly, to try not to micromanage them.

Leaders who have come up through the company as it has grown and succeeded often don't feel like they're different from who they were when they started. They don't realize how scary they've become to others at the company. If the leader doesn't recognize that their role is no longer to oversee a specific function, they likely don't realize how intimidating it is for an employee to tell them to back off — in the nicest way possible. In fact, oftentimes employees won't say anything for fear of negatively affecting their job.

If the leader doesn't recognize that their role is no longer to oversee a specific function, they likely don't realize how intimidating it is for an employee to tell them to back off — in the nicest way possible.

On the flip side, if an executive is hired from a big company, it's the exact opposite. They know they're powerful when they walk in the door, because that's what they've been hired to be. CEOs need to manage not only their own approachability but also that of their executive team.

All kinds of roles and relationships will be changing as you scale. When the company was young and scrappy, the founder CEO likely outsourced back-office functions to friends or cheapest alternatives. *(I know a great accountant from the gym.)* But when they finally hire a CFO or general counsel, there's often uncomfortable tension about changing vendors. A new CFO was hired by a company one of us was advising, and when she started, she discovered that the CEO had been using his mother's accounting firm as their outside advisors. Imagine going to this CEO and telling him that the company's needs had changed, and it was time to fire his mother's accounting firm. She eventually had the conversation, but it took a while. It was tough. He fired his mom.

Having other co-founders on the executive team can also be challenging as you scale, especially if they aren't operating at an executive level at that stage of the company's growth. If they don't grow into their new roles as the company scales, then it'll create tension with the other team members. This is one of those hard conversations that will likely need to take place as you face the Scale Wave.

As the founder CEO builds out their executive team, they may feel lost themselves, wondering what their role is now that every function has its own owner. The move from a founder to a CEO with an executive team is a transformation that isn't discussed enough. To navigate it well requires the founder to move from managing to leading, from doing the day-to-day work to doing the work today that grows the company in three years. If that transformation doesn't happen, some of the most expensive hires in your company will be less effective and productive, and they'll also be more likely to leave.

Not all co-founders can grow into executive level roles as the company scales. Not addressing this early will create tension with the rest of your executive team.

Hard Decisions @Scale

As you're facing the deluge of scaling tasks in front of you, you might feel like you're making a million decisions every day. This is how many entrepreneurs go wrong in decision making, and why they do it.

How Entrepreneurs Screw It Up

- ✗ They don't make decisions. But decisions do get made — they get made by default because time goes by, and the default outcome comes to pass.

- ✗ They make decisions. But they're bad ones.

- ✗ They make decisions. And then they change them — often.

✖ They make decisions. And they are good ones. But the good decision is never (completely) executed upon — effectively making the decision academic — which is also known as a bad decision, because it fails to have an impact.

Why Entrepreneurs Screw It Up

> Decisions as you start to scale have bigger consequences — perhaps millions of dollars of consequences instead of a couple thousand dollars, or perhaps a blow to your brand that could be caught in a bad press cycle. Entrepreneurs who haven't had the experience of making these large impact decisions before will avoid making them, or they'll make them and then change their minds.

> As you conquered your Launch and Pivot Waves, you involved many of your employees in decision making. Now that you have 150 people or more, you're still in the habit of getting lots of people around the table. Or perhaps people push their way to the table because they've always had a seat in the past. You start developing a consensus-driven culture. With 150 opinions, you can't achieve consensus — and the flow of decisions slows to a trickle. You start developing a "large meeting" culture, which eventually sucks up so many people's time that no one has time during the day to actually do the work.

> You finally have a full-blown, experienced executive team (yay!). Fortunately, or unfortunately, this team comes with some big personalities and strong opinions. These folks are used to being heard and expect their opinions to carry weight. The discussions never end. Decisions can't get made.

> You're used to calling all the shots. Now you question every decision everyone else makes. People stop making them.

They lose their agency. You're the bottleneck. But the wave keeps coming at you.

How Not to Screw It Up

If you're going to successfully manage at scale, you must learn how to effectively make hard decisions.

✓ Bound the decision box. Decisions that impact beyond twelve months aren't decisions; they're forecasts or models or goals. Decisions that impact the next twelve months should be decisions that don't change unless something significant has changed in your operating environment.

✓ Balance your analysis. Give your team time to conduct the appropriate amount of research and data analysis and present their recommendation before making a decision. But don't fall prey to analysis paralysis. If your tendency is to make quick, perhaps rushed decisions, then collaborate with a trusted colleague who tends more towards well-researched decisions. If your tendency is to analyze every cost to the cent before being able to move forward with a project, then bring in someone who trends toward an intuitive approach.

✓ Identify decision owners. Give one person (the right person) authority to make the decision. Communicate who that is to the team. Help the team understand that they provide input, and they're welcome to influence, but the decision maker's decision will be the company's decision.

✓ "Agree and commit, disagree and commit, or get out of the way."[6] This quote is often attributed to Andy Grove, former CEO of Intel. Make this your mantra. Realize that sometimes it's you who needs to get out of the way.

---

6 "Disagree and commit," *50 Folds* (blog).

✓ Hold decision makers accountable. With great power comes great responsibility. Make sure the decision maker understands they're responsible for implementing the decision successfully, not just for making it. That includes you if you're the decision maker.

All of this means that in some circumstances, you're going to have to take better control, and in many circumstances, you're going to have to give up control. You're no longer the sole driver of your culture, and you can no longer be the sole decision maker.

Your primary job now is to focus on leading. This means leaving the managing to your managers while you drive focus toward the priorities of the company. It means uncovering existing single points of failure. It means being prepared to proactively address conflicts among groups in your organization. And it means bringing your energy and enthusiasm to keep your team motivated as your company does the hard work of scaling.

Key Takeaways

1. Be purposeful in how you choose to motivate your employees. A clear message might mean that some employees choose not to stick around. But it will motivate the people you want to lead.

2. Great scale leaders limit change, prioritize just a few new strategic opportunities at a time, and complete or abandon one strategic opportunity before adding a new one for the business to implement.

3. Real scale leaders help captain the whole ship rather than worrying about their own little life raft.

4. Scaling requires removing single points of failure (SPOFs) — and doing that while making the SPOF part of finding the solution.

5. You can't just keep adding more chairs to the room, or your meetings will start achieving the productivity of many governments. Set expectations early that these seats aren't tenured positions.

6. The move from a founder to a CEO with an executive team is a transformation that requires the founder to move from managing to leading, from doing the day-to-day work to doing the work today that grows the company in three years.

Sidebar:
Getting Sidetracked on Culture

There's an awful lot of crap you're in danger of ingesting by reading advice about company culture on the internet, or perhaps by paying for it from an HR consultant. Advice like: "First define your company culture. Next, ensure that the culture you have defined is clearly communicated to every new employee who walks through your door."

One of the bigger mistakes a scaling company can make is working on culture statements and then telling everybody about what those culture statements are. Culture statements are fine — if they're the end result of a realistic analysis of the current behavior of the company's employee base. They're a waste of time if they're the starting point of defining your organization's culture. Aspirational declarations about how we act are unlikely to be accurate and reciting them does not make them so.

Culture isn't something you can fix. It's what you get after you fix whatever else is broken. Building a culture is what you and your managers are doing every day to create a sense of common purpose and to broadcast that sense of purpose throughout the entire organization. Sometimes, the most important way to fix your culture is to unburden your enterprise of whatever's dragging it down.

Mistake 12: Staying Within the Four Walls of Your Own Product

Scaling your product and your team are all in service of the ultimate scale: growing your customer base so you can grow your revenue and profits. This requires your scaling company to do two things at the same time: continue to retain and grow your early adopters while also appealing to the early majority. Your next round of customers, the early majority, have different characteristics and are driven by different motivations than your original early adopter customers.

You're used to selling to customers who are focused on the possibilities of what your product could be. These early adopters care about how your offering can address their problems today and in the future. Your early majority prospects are focused on today. They aren't interested in abstract discussions about what could be. They'll ask you questions about how your product works — and what they really mean (whether they say it out loud or not) is how it works for them — today — in their workflow and daily routines. See our **Sidebar: The Right Customers for Each Wave** for more on the characteristics of each customer type.

As you grow your company and try to scale beyond the early adopters, you'll need to become better at understanding how your customers are using your product within their particular operating environment. This will help you become not just an expert on the four walls of your product, but in helping your bottom-line, practical, early majority customers get full value from your product across their users.

The early majority have different characteristics and are driven by different motivations than your original early adopter customers. They'll want to know how it works for them, today, in their workflow and daily routines.

From Selling a Product to Selling a Customer's Journey

It's often difficult for leaders to see things from their customers' perspectives. And yet it's so easy to see a company's shortcomings when we're the customer, struggling to get our vendor to help us make the best use of their product. Who hasn't had a frustrating customer experience with an airline, or a plumber, or a SaaS vendor who wants to sell you new features — instead of really taking the time to understand how you're using their product, and how they can help you realize value from what they've already sold you?

To manage customer retention effectively, it's imperative companies develop their understanding of their customers' journeys. We'll talk more about customer journeys throughout this chapter.

Mona recalls a team at one of her companies trying to deploy a business intelligence solution across the company. The vendor offering the solution was a mid-stage startup in its early scale journey. In the sales pitch, the vendor's sales team went on about how great the product was and how the vendor had been steadily growing its customer base. The sales team also pitched their professional sales team, who would help configure and implement the solution, so it met the company's specific needs — for an additional price.

The vendor's sales team had done a great job finding a sponsor within the company and worked with that sponsor to close the deal.

But nine months later, no one other than the sponsor and his small team was using the solution. And they weren't using it that often. With only three months left in the one-year subscription, the vendor appeared to realize they were in danger of losing the company as a customer. So they pulled together an account management team and scheduled a series of meetings with the sponsor to try to get him to drive adoption across other teams.

The sponsor was still enough of a fan that they brought the heads of some of the other teams at the company who they thought could benefit from adopting this business intelligence tool. They all attended a meeting in which the vendor went through a sales deck all over again, pointing out the product's features and showing quotes from other happy customers.

(As an aside, most companies that haven't learned how to conquer the Scale Wave default to selling "functions and features." But during the Scale Wave, your new customers are looking for your guidance on how they can use your product within their company, which goes well beyond a recitation of the features of the product alone.)

The company didn't get much help from the vendor in understanding how to best customize the product to fit within their internal processes. So they tried doing it themselves. A cross-functional team spent about a month coming up with convoluted workflows to try to fit this solution into their customized way of doing things. They weren't the experts in business intelligence deployment, so there was only so much anyone could expect from them.

The resulting workflows were handed over to the vendor's account management team who lobbed it over to the vendor's professional services team. Well, the entire lifeblood of a professional services team is to do custom services work. And this services team was no different. They spent two weeks working on a statement of work and came back with a substantial management consulting proposal — an effort that was larger than the value the solution provided in the first place. Needless to say, the sponsor didn't get approval to renew the solution for another year.

What went wrong? The vendor didn't have a clear vision for how to drive adoption within the company's environment. Their customer-facing teams were all operating within the four walls of their own product, and not from the perspective of the journey their customer had to take when deploying their solution. In the end, they wasted time with an expensive, untested, resource-heavy service offering that wasn't reflective of the value their product had been delivering so far. The company was looking for a clear and immediate ROI, but the vendor gave them the opposite.

Mona spent some time researching some underlying metrics of this vendor. They were adding new customers, but the undertows were in their retention and payback metrics. They were lower than industry average, meaning that not enough existing customers were renewing their contracts with them.

In the world of software, it costs around five times more to acquire a new customer than it does to retain an existing customer. And research by Fred Reichheld, who pioneered the Net Promoter System, suggests that by increasing customer retention rates by 5 percent, a company can increase its profits anywhere from 25 to 95 percent.[7]

> In the world of software, it costs around five times more to acquire a new customer than it does to retain an existing customer.

The Importance of Post-Sale Touchpoints

What are customer journeys, and why are they important as you scale your business? Search "customer journey" on the internet, and you'll be bombarded with the following five-step process.

---

7 Reichheld, "Prescription for Cutting," Bain & Company.

THE 5 STAGES OF THE CUSTOMER JOURNEY

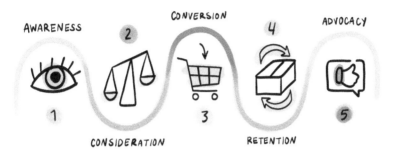

And yet most articles about customer journeys focus on understanding the steps your customers take before buying your product and becoming a customer (the first three steps of the Customer Journey in the image above). Understanding these steps helps you tailor specific marketing campaigns to prospects depending on where they are in their journey toward purchasing.

Yet when you're scaling your company, it's just as important to understand your customer journeys after they buy your product, the Retention and Advocacy steps. This is especially the case in businesses where it costs more to acquire a customer than the revenue you get from your first sale to that customer — which is almost always the case in SaaS businesses — because your business is reliant upon the customer continuing to buy from you to generate a profit. And only happy customers renew their business with companies.

When talking about customer journeys, marketers constantly focus on understanding or mapping the touchpoints or interactions a customer has with your company, such as the purchase touchpoint, or a customer support call interaction. They focus on how the customer feels at each of these interactions.

In our view, this is much too narrow of a perspective on your customer's journey. Because 90 percent of the interactions a customer is having with your product have no touchpoint to your company. And by the time they call you for customer support, they're already, by definition, not thriving.

> In the customer's journey, 90 percent of the interactions they're having with your product have no touchpoint to your company.

Let's look at an example of a real post-sale customer voyage in which internal touchpoints were ignored to the company's detriment. A company purchased a contract management system and worked with the vendor to deploy it in their environment. Then they spent two months without the vendor's help uploading existing contracts into the system. (Here's the first set of internal touchpoints the vendor didn't have on its radar, because they weren't involved in the company's deployment efforts.)

Two months later, the company decided to run an internal audit to see how well their new contract management system was performing. Their audit revealed two things:

1. They hadn't managed to upload all their contracts and weren't continuing to upload new ones.

2. Many of the "automatic" tags, which were supposed to be a great feature of the system, didn't generate the correct information. The contracts that were uploaded were getting tagged improperly by the system. (This is the second set of internal touchpoints the vendor didn't have on its radar.)

The company called vendor support and logged a ticket. (First external touchpoint with the vendor since the deployment.)

Meanwhile, they tried to build their own internal process for getting various people to upload contracts to the system instead of just saving them on their hard drives or in Google Drive. Some people remembered, some didn't.

That's when they decided to build a custom integration between the contract management system and other systems, so contracts were uploaded and routed automatically. Their team created training materials for internal teams to teach them how to use the system.

Vendor support finally responded to their ticket and passed them on to the professional services team. They scheduled a call and explained the problem to the professional services team, and they made some changes. Their advice was to implement some of the other features the company hadn't made use of yet. The company never got around to figuring out what those additional features did. It seemed very complicated. And so, they never implemented them.

Months later, they tried to find a particular contract in the system, only to discover that the internally built integration had broken. Contracts had not been uploaded for months!

After eleven months, they received a call from the contract management system account rep telling them that it was time to renew their contract. Obviously, that didn't happen.

You may notice that out of these many steps in the customer journey, only a handful of them were interactions with the vendor. And sadly, even those few steps weren't handled well by the vendor.

This vendor thought they were in the business of selling a product. In actuality, the customer was an early majority company trying to deploy a solution that worked seamlessly in their operating environment, so they could get immediate value from the purchase. What they needed was a vendor who:

✓ Understood that to be valuable, this product needed to easily and efficiently store and organize contracts that were being generated in multiple locations within the company

✓ Realized that early majority customers weren't going to spend a great deal of time figuring out the complex functionality of a new product, and that coming in with a plan to help them gradually ramp up familiarity with the product features would improve their usage

✓ Delivered training materials instead of leaving it to customers to develop it themselves

✓ Offered integrations into the most common systems customers used, so they didn't have to build half-baked integrations themselves

Looking at this from this customer's perspective, it's obvious that the vendor under-invested in the "retention" step of our customer journey — and it cost them an account.

Making Your Users Heroes

Riding the Scale Wave means learning how to "value sell" throughout the first year of your customers' experience with your product. The difference between value selling before the deal closes and after is who you're selling to. Often before the deal closes, you're selling to the "buyer" at your customer — the person who owns the budget and has the authority to spend the money. That person cares about whether this expense will help them grow revenue, cut costs, reduce risk, or a combination of these.

After a deal closes, you're "selling" to your users — the people who must engage with your product. What they care about is that you're making their lives easier, and even better, you're making them look like heroes. The "value" is different before and after a sale. Companies that don't invest in post-sale value selling often fail to scale because too many of their customers churn.

> Companies that don't invest in post-sale value selling often fail to scale because too many of their customers churn.

So how do you make your users' lives easier and make them heroes within their organizations?

✓ **Develop journey maps and common use cases.** Engage in meaningful conversations with customers who share similar personas and customer profiles to develop common use cases. This allows you to tailor journey maps to the needs of each customer profile.

✓ **Chart customer journeys.** Craft comprehensive customer journey maps for each customer profile. By documenting their interactions with your solution, you gain a deeper understanding of the outcomes your customers are trying to drive with your products and services. Not every customer will have the same journey map. But often enough, customers within the same segment (e.g., healthcare companies, e-commerce companies, manufacturing companies) will have similar enough environments that a successful scaling business can develop a generalized journey map template that drives value from using the product within each segment. And then each segment template can be tweaked for each customer.

✓ **Enhance internal product touchpoints.** Go beyond customer-company interactions and establish best practices for internal touchpoints that your customer has with your product. Helping your users streamline their internal procedures will turn them into heroes.

✓ **Simplify customer journeys.** Constantly work to simplify your journey map templates. Your customers don't have time to learn and navigate complex processes on their own. Recognize that your journey maps might require a customer to change their current work practices a little bit. Try to limit required changes, but don't be afraid to recommend changes as a best practice — especially if you know it's going to show immediate value to them.

✓ Deliver useful training and training materials. No one remembers a new process from just one training session. Build training directly into your product to achieve scale and accompany it with training materials that empower customers to train others. Doing this helps drive broader adoption across an organization.

✓ Deliver seamless integrations. Integrate your product into the broader customer systems and environment, enhancing its value within the customer's established workflow.

✓ Foster product-driven efficiency. Over time, "productize" these workflows, best practices, and training directly into your product. This approach not only simplifies user experiences, but also drives scalable efficiencies.

✓ Provide accessible "on-ramp" support. Offer accessible office hours during the initial months for each new customer to make sure they're successfully engaging with your product in the critical early stages.

✓ Practice continuous success monitoring. Commit to ongoing success monitoring to proactively address any issues and optimize customer experiences in real-time.

✓ Create transparent reporting. Provide transparent reporting to your executive management, demonstrating the positive impact of your efforts on customer satisfaction and the company's bottom line.

Your Customer Success Team

You're moving from selling to the pioneers — those spry early adopters who like to try new things — to the larger, more established companies that already have complicated workflow systems.

They don't have time to bring on a standalone product that doesn't fit how they work.

To do all of this, you need to invest in a different customer success team from your Launch Wave days. Customer success is now much more than support, and more than onboarding (although if you don't do onboarding well, you're toast from the start). When you fail to train your customer success team well, everyone defaults to generic responses and focuses on features and functions. If your success team isn't engaged, your customers aren't engaged. What you're really doing is selling the product and leaving it up to the customer to figure out the best way to extract value from it.

 If your customer success team isn't engaged in figuring out how your customer extracts value from your product, then your customer won't be engaged.

Success teams are only really successful if they're trained to think outside the four walls of your product and to extend their understanding to a variety of customer-specific environments. When they aren't trained that way, they tend to become either firefighters (see **Mistake 10: Staying in a Constant State of Firefighting**) or order-takers. To compensate, companies tend to just throw more people at the problem once they see a customer isn't engaging with the product. They'll insert a salesperson, a solutions consultant, someone from professional services or another product, or a customer education/training team into the process.

You could have great products, speedy delivery, and a dedicated customer service team, but any weak link in what turns out to be a very long chain could send your customers elsewhere. And adding more and more resources to solve your problem is definitionally the opposite of scaling.

There are many elements you need to master to scale your customer base. An entire book could be written on this topic. In this chapter, we've focused on keeping your customers after you close the first deal with them. Why? Because to scale profitably, particularly in SaaS, you need customers to come back and buy from you again and again. As you work through this successfully, you'll build a base of happy, long-term customers. The next step is getting those customers to spend even more with you.

Key Takeaways

---

1 Scaling requires your company to both continue to retain and grow your early adopters while also appealing to your next round of customers, the early majority.

---

2 Scale leaders help their customer-facing teams move from operating within the four walls of their own product to understanding the journey their customers take as they use the product within their business environments.

---

3 It costs approximately five times more to acquire a new software customer than it does to retain an existing customer. Companies that don't invest in post-sale value-selling often fail to scale because too many of their customers churn.

Mistake 13: You Treat Your Customers as Nothing More than Customers

In the previous chapter, we considered the mistakes scaling businesses make in the first post-sales stage of Retention. Building an organization that's excellent at retaining existing customers is one of the key skills of a great scale leader. You can't conquer the Scale Wave if you keep losing customers that cost you so much to acquire. The Scale Wave is a long one though. And the next part of that wave will threaten you in a different way.

As your customer base grows from ten to one hundred to one thousand and more, your company will spend more and more on every part of your business to keep up with the growing numbers. If your spend-to-revenue ratio isn't getting smaller, then you aren't improving your bottom line — your profitability — and you aren't scaling. Real scaling means that you've figured out how to spend less to get the same amount of revenue.

If your spend-to-revenue ratio isn't getting smaller, you aren't scaling.

Scale leaders learn to get creative to reduce their sales and marketing spend while still growing their customer base. Mistakes 13 and 14 focus on two programs successful companies develop to do just that.

Moving from Users to Humans

The first program to help you scale sales and marketing requires you to stop treating your customers as nothing more than customers.

It costs a lot to maintain customers, and even more in marketing and sales spend to get new ones. And that spend is often wasted because customers ignore your marketing campaigns, complain about your customer support, demand more features, and still insist on bigger discounts.

But if you can turn your customers into advocates of your product, both within their own companies and in the larger user community, your customer advocates can do your marketing for you. To do this, you must first give your customers more value than just *allowing* them to buy your product. You must treat your customers as more than just customers.

> Turn your customers into advocates of your product, and they will do your marketing for you.

Turning Customers into Communities

Jennifer Susinski, a customer advocacy and community manager who has grown customer communities with a few hundred members to ones with many thousands, learned all this the hard way — her customers told her.

Jennifer didn't spend her entire career in customer advocacy — she got thrown into it. A company she was working for wanted to improve their customer satisfaction rating. When a company starts investing in their customer satisfaction score, it's usually a sign of some underlying problem it's trying to solve. Jennifer didn't know what prompted this investment at the time, but she agreed to take on responsibility for the customer community. She started with a listening tour to better understand how customers in the community felt.

The feedback was painful. Customers described the community as "a condescending one-way street." They were looking for support, and when they weren't getting it from the customer support team, they would try the customer community. But the company had relegated community management to a third party whose main strategy was to feed existing online and email content through to the community. Not getting what they needed from the support team — and feeling patronized within the community — was more than enough to impact overall customer satisfaction.

Why was Jennifer being asked to turn the community around now, after years of relegating it to a third party to manage? She didn't know the answer at the time, but later Jennifer learned the company had been in acquisition talks. Proving the company could scale the next wave was likely a critical ingredient in those discussions. And an effective customer community is one of the best ways to prove scalability exists.

Jennifer examined the source of the problem — she began by tracking every issue sent to customer support. As with many companies struggling to scale, her company's support organization wasn't set up for success. From the outside, it looked like support requests got bounced around from one person to another without sufficient attention to the customer or resolution. It's easy to blame the support specialists, or perhaps the leader, the systems, or a number of other individual elements within the organization. In reality, this is a failure to ride the Scale Wave.

Jennifer didn't run the support team. She couldn't change what was happening there. But she realized she could help keep the rising water from sinking the company's boat by fixing the customer community.

Jennifer kicked off her project by focusing on three things:

1. Helping leadership see a customer support issue as human pain

2. Getting to know the customers in the community personally

3. Empowering customers to get to know one another

Jennifer uncovered one support issue that was particularly troublesome. A hospital was having problems using the company's product, which was affecting the hospital's ability to deliver strong service. Jennifer engineered a meeting between her company's product general manager and a critical user from the customer. Initially, the general manager tried to do what we all default to doing. He asked to bring his team to the meeting. Jennifer insisted he come without his team. Her goal wasn't for everyone to sit around troubleshooting a support issue. It was to get the GM to see his customers as humans.

To say the GM was embarrassed after hearing the customer's story is an understatement. He not only set out to solve the product issue for that customer (and others), but he also agreed to give the customer a free subscription period in recognition of how the company had failed the customer so far.

Now having a one-on-one with every customer, and giving away product, isn't a successful path to scale. But that wasn't Jennifer's goal.

It's hard for a startup to focus on its first set of customers after they close the deal because its growth is so dependent on securing the next set of customers. Imagine how hard it is to really pay attention to your existing customers when you have hundreds — or thousands? As Jennifer's company was trying to show it could scale, it wasn't just drowning in customer support calls. It suddenly had other challenges that didn't exist when it was smaller — employee relations issues, internal systems that needed upgrading, performance management, performance improvement plans, training needs, market pressures . . . Sometimes there's so much going on, it's hard to remember that the point of it all is happy customers.

> As Jennifer's company was trying to show it could scale, it suddenly had challenges that didn't exist when it was smaller.
> Sometimes there's so much going on, it's hard to remember that the point of it all is happy customers.

With her GM's newly gained support, Jennifer started scaling customer satisfaction through community advocacy. She says, at its peak, this community had nearly 5,000 users. And she knew something personal about at least half of them. Her best practices included the following strategies:

✓ She made the community a "safe space" for customers. That meant that sales and account reps weren't part of the community, because that would make community participants feel like the forum was just another sales and marketing machine.

✓ She created customized customer contests. These contests helped customers open up about themselves as well as their product and business challenges.

✓ She launched a monthly "deep dive." In these sessions, she asked everyone to share two fun facts about themselves and one issue or piece of advice related to the product that others could learn from.

✓ She cared enough about what she learned about each customer to engage with them on a human level. She found herself sharing recipes with one user and sending personalized notes to others.

This hard work allowed Jennifer to know the customers in the community personally and empowered customers to get to know one another. And it set up the foundation for scale.

Great Customer Communities Convert Customers to Fans

Jennifer understood that to reap the benefits of a great customer community, the community has to drive the conversation, not the company. That's one of the most common mistakes companies make when they try to build customer communities. When a customer posted a problem or challenge in Jennifer's community, instead of answering it herself, Jennifer would tag in another customer she knew was likely to have some relevant advice. She knew who to tag because she'd invested so much in getting to know a large portion of the community on a human level. This drove customers to talk to one another and lessened the perception that the community was a "one-way street" of communication. Jennifer recalls four customers who discovered they all lived within 25 miles of one another and ended up meeting at one of their houses for pizza.

> To reap the benefits of a great customer community, the community has to drive the conversation, not the company.

As customers began responding to each other, support issues started to decrease, and customer satisfaction increased. Yes, the result was a more scalable and efficient way to support customers. That's important. But far more valuable is the fact that these customers developed a stronger loyalty to the company.

The community of customers became advocates of the company and its products because they knew someone was paying attention to them and trying to help. People are interesting. Often, they develop a sense of gratitude when they see you trying to help, even if, in the end, you aren't able to. Jennifer recognized this. She would ask, "What can I do to help you? I might not be able to solve your problem, but I can get you into the right swim lane."

Customer loyalty is the pinnacle of a company's desire. As Jennifer says, loyal customers give you more room to make mistakes. When your company hits a bump in the road, your loyal customers will stick with you and even help you work through it.

By being human with your customers, you convert them from customers to fans. And fans are critical in driving the external energy you need to conquer the Scale Wave.

> By being human with your customers, you convert them from customers to fans.

It Takes a Small Village to Build a Bigger Village

There are also some things to avoid when you're building a customer community, according to Dave Hansen, an award-winning expert in customer marketing and client advocacy who has rebuilt two customer communities to become successful advocacy groups during his career. One of the companies Dave helped had grown so quickly they couldn't keep up as the Scale Wave hit them. Dave looked at the customer community as a way to help the company navigate this Scale Wave.

The community had stagnated. Dave's assessment uncovered a number of problems, each of which arose from a lack of understanding about the power of community.

✖ Like Jennifer's company, Dave's company had assigned only one less-experienced person to run the community. The company appeared to view the community manager role as more of a system administrator than a strategic role to help the company conquer the next wave of its growth.

✘ As a result of this choice, no one else in the company was involved in the customer community, either directly or through accessing feedback from it.

✘ A successful customer community is a treasure trove of data for a company. But this company hadn't built any analysis into the community platform.

Getting customers to join your customer community is a bit like getting your employees to answer your employee engagement survey. If you make them do it, and then nothing comes of it, it's worse than if you didn't do it to begin with. Running a customer community that offers a poor experience decreases your overall customer satisfaction.

> Getting customers to join your customer community is a bit like getting your employees to answer your employee engagement survey. If you make them do it, and then nothing comes of it, it's worse than if you didn't do it to begin with.

Dave set out to change things. He decided to start by creating more ownership of the customer community across the organization. He scheduled regular "pacing calls" with different internal departments to understand how they were pacing against capabilities and outcomes the community was asking about. Each call included a "Voice of the Customer" section in which he shared what he was hearing from the community.

Dave also insisted the community could not be managed by just one person. That would create a SPOF (a single point of failure — an issue we discussed in **Mistake 11: Managing Like a Startup Founder**). He pulled in resources from different parts of the company to expand ownership.

He then implemented better systems and analytics to be able to mine the data collected from the community. He owned all the customer sentiment analysis and regularly shared it across teams, so it could inform roadmaps. Eventually, there were dozens of people across the company who were regularly benefiting from information collected from the community.

Many companies create ineffective customer advisory boards. Dave didn't want to waste his customers' time. Creating more ownership across the company enabled him to show his customer advisory boards that when they provided feedback, the company considered it thoughtfully. Even if it wasn't acted upon, Dave could show the thinking and reasoning behind the company's decision.

Dave's smaller internal and external communities helped create a flywheel for the broadest of the company's customer communities. He built little villages that helped give energy to a large and involved community.

Beyond a Brand

As you scale, you're going to have to make room in your organization to actually do the work of building a sense of community among your customers. Buying some community software and configuring it to send out generic emails periodically to your customers isn't just a waste of time — it's condescending and banal. To the customers, it feels like just another way to send marketing messages they'll inevitably delete.

Building a thriving customer community takes work, resources, and people paying close attention. You must build trust — and you do that by realizing your customers aren't just users, they're also human. It takes time to get your internal teams to pay attention to them in a different way from the structured channels your company put in place at the beginning of its scaling journey. And it takes work to get your customers into a cadence of talking to each other, so eventually, they're driving the conversation instead of you. When customers are helping others in the community, they develop a sense of loyalty that galvanizes the overall business through the Scale Wave.

> You must build trust — and you do that by realizing your customers aren't just users, they're also human.

The best scale leaders understand the value of investing this time and effort. Because great communities are better than brands. A great brand can sometimes create communities, but great communities more often make the brand.

Key Takeaways

1. If you can turn your customers into advocates of your product, both within their own companies and in the larger user community, your customer advocates can do your marketing for you.

2. To reap the benefits of a great customer community, scale leaders let their customer community, not the company, drive the conversation.

3. Running a customer community that offers a poor experience decreases your overall customer satisfaction.

4. Scaling through customer communities requires companies to invest in building trust — and you do that by realizing your customers aren't just users, they're also human.

Mistake 14: Partnering as a Side Hustle

Few companies manage to create a successful partnership ecosystem, yet a 2020 Ernst & Young survey found that 68 percent of corporate business leaders viewed partnerships as a critical way to succeed in the market.[8] Building a successful partnership program isn't a side hustle. It's like launching your second product. It takes the same amount of experimentation and upfront investment, and roughly about the same length of time before you see the fruits of your labor. You'll hit the same waves: a Launch Wave, likely a Pivot Wave, and then a Scale Wave.

Most businesses make the mistake of underestimating the effort it takes to launch a partner program that returns value to the business. So why go through all of this again? Because a successful partnership program can help you solve important scaling challenges you'll face:

✖ You aren't Where Our Customers Live (WOCL).

✖ Your lack of post-sales services is hurting your business.

✖ You're missing an opportunity to sell to other markets.

✖ You're missing an opportunity to turn a potential competitor into a partner.

✖ You don't want a partner program, you just want a partner.

Challenge 1: You Aren't WOCL

People who think they're smarter than you will tell you your product has to be used every day before you can retain customers

---

8 "How to drive innovation," EY.

and scale. They'll ask for your daily active user (DAU) metrics. We don't think those people are that smart. What's more important is that your customers stick to using your product, no matter what the natural usage cadence is. What's important is that your product is sticky.

Stickiness means that once someone incorporates your product into their world and starts using it, it becomes very hard for them to remove it. Some products are sticky because they're designed to be habit forming — think Instagram or TikTok. It's much harder to make a B2B product habit-forming. In fact, users of many B2B products feel rather indifferent, if not actively resistant, to using them.

To make a B2B product sticky, it has to be an integrated part of your customers' workflows and operating cadences. Our term for this is being WOCL — being Where Our Customers Live.

WOCL is our term for being Where Our Customers Live — by integrating our product into our customers' workflows and operating cadences.

Let's say you're selling software that lets you scan a receipt using an app. The app automatically reads the receipt and categorizes it. Then it submits the receipt to your accounting department in the format the accounting system can understand. There are many of these kinds of business apps, and they've been around for over a decade. And yet many companies still manually send their receipts to admins — humans — to have them manually record them in the company's systems.

At its core, this is because remembering to use the app is a habit that must be formed. And for many reasons, possibly because you don't have to expense business purchases every day, many people don't form the habit of opening the app. The largest group of people who are authorized to expense their purchases already have a habit — they send their receipts to their admins.

If you're a leader at such an expensing app company, you may be influenced by those aforementioned "smart" people to offer more features in your app to try and get people to use it more often. Have fun with that. New features should be driven by customer needs, not by your company's failure to scale.

> New features should be driven by customer needs, not by your company's failure to scale.

Instead, what if you go WOCL? Where are your customers every day? What software are they using every day that's connected to purchases? What are they purchasing that they can expense to their employer? Find those places and integrate your app into them.

What if, as an example, when we use an app to order a rideshare, we could press a button in the app and the receipt would automatically be sent to our accounting department for reimbursement? Your expensing app company could go to where your customers live and build an integration into these ridesharing apps. Because we've already developed the habit of using those apps. Meeting us there with one simple button would have us quickly adopting your app. And once we do that, we're less likely to take steps to integrate anyone else's expensing app into our ridesharing apps. You've inserted yourself into another app's ecosystem — a set of interdependent applications that takes more work to dismantle than it's worth. You've achieved stickiness. (As an aside, Uber has done exactly this for expense management apps by creating interfaces for these apps to integrate into Uber's system. And a few expense management apps have taken advantage of it.)

Okay, but what does this have to do with building partnerships? Well, it's unlikely that your development team can integrate your app into a ridesharing app on its own. They'll need some assistance, and perhaps co-development from the ridesharing companies' development teams. But they are developers. It's not their job to reach out and try and convince other companies — that are often much larger than yours — to work with them. In fact, that's your job. And it involves building a technology partnership with the companies whose ecosystems you should be a part of to achieve the scale you desire.

Challenge 2: Lacking Post-Sales Support

Perhaps you've built that B2B product that's so well-designed and so intuitively integrates into your user's regular workflow that every user feels like the character John Anderton in *Minority Report* (2002), elegantly swiping and twirling their hands to right corporate wrongs through the power of the product you've delivered. If that's you, skip this section — you don't need post-sales services.

For the rest of us, while our products are still valuable, they often need to be paired with strong customer onboarding and adoption services for your customers to fully understand how to enjoy your product's benefits broadly across the organization. But like everything we must do during the scale stage, building an effective, scalable, global after-sale services organization takes resources, people, time, and commitment, all of which are in very short supply in your scaling organization.

Enter services partnerships. Building successful services partnerships with other companies probably isn't going to save you time at the start. Because partnerships aren't a side hustle. But they will help you scale faster in the long run.

Services partners come in two different flavors: 1) smaller companies that provide local or industry-specific support services for your product, and 2) huge companies that deliver consulting services to companies and can build multi-million-dollar service offerings that include deployment of your product. Both are valuable, and both have their advantages and challenges.

Building an ecosystem of smaller services companies that deliver great services to your customers can have some pleasantly unexpected outcomes. If you can give a small services company a reasonable number of your customers for them to service, then they'll develop an expertise around your product and be motivated to encourage more companies to use your product, so they can expand their own service client list. Creating a symbiotic broad-based ecosystem of small service provider partners is a great way to accelerate your company's scaling efforts. A Salesforce study revealed that by 2026, for every $1.00 Salesforce makes, its partner ecosystem will make $6.19.[9]

> If you can give a small services company a reasonable number of your customers for them to service, then they'll develop an expertise around your product and be motivated to encourage more companies to use your product, so they can expand their own service client list.

On the other end of the spectrum is partnering with a large, often brand name consulting company. These types of partners (sometimes called system integrators) build out complex solutions for their clients by combining one or more hardware and software products from multiple vendors. Accenture, Deloitte, IBM Global Services, and PwC are just a few examples of systems integrators. Because their services are tailored to large, complex deployments, they often only invest in solutions that allow them to charge millions of dollars in associated services. This makes it quite hard for early-stage scaling companies to get the attention of these large system integrators. This is where your investment in understanding your customer journeys starts to pay off.

If you've done your homework and developed an under-standing of your customers' workflows and how your product fits

---

9 Salesforce, "New Study," Salesforce.

into them, then sharing this expertise with a systems integrator so they can build a large-scale services offering around it is an incredible way to break into the lucrative large enterprise market.

One note on building a post-sales services partnership: if you already have an internal services team that's charged with generating revenue by delivering services to your existing customers, then you're in for some channel conflict challenges. They won't appreciate the company's efforts to send service opportunities to your partners because it interferes with their revenue goals. You get what you incentivize. Instead, change your services team's key performance indicators (KPIs) to increasing retention by enabling service partners. If your retention rates increase, the company still gets the benefit of additional revenue, while having removed the channel conflict that's bound to stall out even the best partnership teams.

Challenge 3: Missing Market Opportunities

Let's say you've launched your business selling your product to schools and universities. Every now and then, your sales team gets an inbound lead from conference organizations wanting to see if your product might work for their needs. Because you have a great sales team, they'll follow every lead. But they need different marketing collateral. And they need different data about this new market segment. They need post-sales support that's geared to conference organizers rather than students. Sadly, no one is delivering on their requests — because every team is in firefighting mode or is already overstretched investing in scaling.

Instead of throwing out what might be a great additional segment to help your company grow, what if you found a partner who already sells products to conference organizers? They already know how to market to and support that group of customers. You're still going to have to train them to sell your product. But you don't have to execute the sale or the post-sales support.

A less obvious, but more common, strategy is to choose partnerships when deciding to grow sales in other countries. It's tempting for your sales team to want to just build a new sales hub in another country. Inexperienced sales leaders will believe it's just a matter of hiring local salespeople and starting to hunt for business. Of course, nothing about scaling is that simple. Among the many things you'll have to manage are local marketing conventions, 24/7 global support, local employment compliance and country-specific regulatory requirements, possible translation requirements if you're selling in a country that speaks a different language, and culture considerations. Partnering with a local, reputable reseller that can take over marketing, sales, and support takes much of this burden off your plate — once you train them.

Challenge 4: Missing Partner Opportunities

A scaling B2B software company was the leader within its narrow market segment. It looked to broaden its market opportunity through partnerships with larger companies that sold into larger but complementary markets. After years of discussions, one of these partners offered them an opportunity to integrate their product's functionality into the partner's solution for the partner to resell under their own brand (aka white labeling).

Implementing this partnership would mean both companies needed significant product development work to create a seamless integration. It also meant that the startup would give up a significant percentage of the selling price attributable to their product to the larger company that was integrating it into their solution.

After all the work of getting this partner to the table, the startup's executive team couldn't get out of the way of its own ego. The finance team didn't want to give up the revenue from selling their own product directly. The sales team felt threatened by having another company's sales team outperform them. Product and engineering already had a long product roadmap and didn't want to change it.

The few executives pushing for this partnership had some good reasons for doing so. The partner had a sales team four times the size of the startup's team. The much higher sales volume the partner would achieve would more than make up for the fees the partner would collect.

In the end, the startup chose not to pursue this partnership opportunity. And within the year, the larger company announced a competitive product they were building internally. The startup wasn't willing to play with them, so the would-be-partner decided to compete.

Challenge 5: Stopping at One Partner

Getting one partner is like running one experiment. As you try to work with your first partner, you'll learn that your scaling company hasn't developed any skills in partnering. Your business has been heads-down focusing on building products, marketing and selling products, and (hopefully) servicing customers. It's a lone boat sailing among an armada.

Going into partnering thinking you can pull off just one successful partner relationship and then leave it at that is a bit like thinking you'll build an MVP and not have to take it any further. It doesn't take you very far given the investment you must make to get it launched.

To see a good ROI on your partnership efforts, you must leverage the investment you make to achieve one successful partnership (which is likely not going to be your first one) into a playbook that'll help you make many more partnerships successful.

While there are certainly businesses that have succeeded entirely due to one critical partnership, that model is not the norm, and comes with its own baggage — because if that partner goes away, it could crash the business.

Common Partner Types

So, we know we don't want just one partner. Let's talk a bit about the different types of partners you could have. You want to think about partners as part of an ecosystem that works together, helping you reach different objectives.

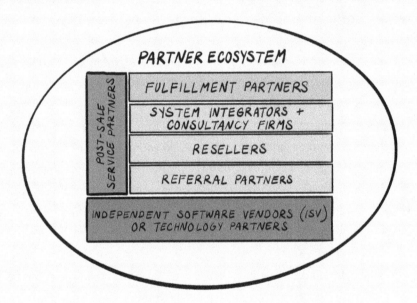

An Independent Software Vendor (ISV) is a technology partner that offers a solution that complements your product or your target user. Ideally, a technical integration is developed between your product and that of the ISV, resulting in a unique joint value proposition. ISVs often help fill a technology gap that enhances the user experience.

To get your product introduced to more customers, you'll want a referral partner. Referral Partners are individual representatives or small firms that represent vendors and introduce them to their network of customers. They earn a fee or margin from each sale they introduce. The vendor account rep owns the sales cycle. These partners refer leads to your direct sales team. The relationship is best structured by agreeing that you'll pay some percentage of the first year's revenue (usually 3 to 10 percent) if the lead is approved and if your direct sales team closes the deal on that lead within a certain period of time.

You may consider a reseller partner in order to get your product out in bulk numbers. Resellers are organizations that buy your products at a discount (VARs, or Value-Added Resellers, will also add value to your product) and then resell it. These partners are committed to sourcing and closing customers on your behalf. You'll have to invest in enabling them and consider channel conflict (meaning limiting their ability to resell to certain territories or markets that aren't already covered by your direct sales team).

If your product is facing the challenge of infiltrating an already established system of products, a system integrator is an advantageous partner. System Integrators (SIs) are large, often well-known consulting or services companies that specialize in coordinating, digitizing, implementing, maintaining, planning, and testing a large and complex service. SIs have deep technology expertise in certain digital transformations and often deliver services in the millions of dollars that combine a few product offerings from vendors (such as your company) along with expert "proprietary" services developed by the SI.

Fulfillment Partners simplify the vendor management process with customers. Your goal is to make it as easy as possible for your sales team to close deals. If enough of your customers would much prefer to purchase from an existing fulfillment partner, it's a great idea to negotiate a fulfillment partner agreement with them.

Customer service can drain a lot of time and resources. Post-Sales Service Partners are strategic services firms that have the expertise — or are willing to invest in the expertise — to learn how to support customers using your product. Normally, you'd commit to training them and certifying them. You'd also revoke their certification if they weren't providing services at the appropriate level.

Yeah, but How Hard Can Building a Partnership Program Be?

It can be hard. And there are many reasons why.

Go-to-market partnerships are likely to scare your direct sales team. They'll feel threatened that someone else might be selling to their customer base. To work with your sales team, you'll want to give them credit (usually through commissions or quota relief) for your partner's sales. But this will have you facing the wrath of your finance team, because you're already paying your partner for the sale, and now you're asking to pay your sales team for that same sale.

Your ops team will apply your existing business KPIs and metrics to this new fledgling partner program you're trying to build. Like any new initiative, your embryonic partner program won't be able to compete with the metrics generated by your existing business. This will cause your ops team to question further funding.

> Like any new initiative, your embryonic partner program won't be able to compete with the metrics generated by your existing business.

You'll need to develop specialized marketing collateral for your partners that's different from what your marketing team normally creates for your customers. But your marketing team is already underfunded and doesn't have resources for your partner program. Your requests will stress out your marketing team.

Despite all these challenges, let's say your small team does manage to bring on half a dozen partners. Now you need partner success support. Technical partnerships require scarce resources to invest in building integrations with your product. Channel partnerships must invest scarce sales resources in understanding the value of your products for their customers and trusting that value can be realized. All partners must invest in training their teams on how your product works and why the partnership drives value. Your partners won't make this investment if you aren't supporting them and providing the materials they need. And, of course, this additional support request won't help you make friends with your success team.

We've already covered how frustrated you'll make your services team if you try to bring on a post-sales services partner. And these are just the internal challenges you'll face. We haven't even started talking about the challenges of convincing other companies, already over-extended on their own internal goals, to spend time building a partnership with you.

You might've thought you could create a nice, neat project plan to get you from start to success with your partner program. Something that looks like this:

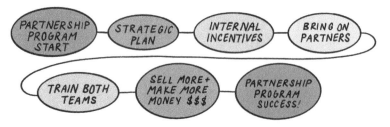

We could pretend we'd be helping you by identifying a bunch of steps under each of the categories in the above picture, but in reality, your partner program efforts are going to feel a lot more like your Launch Wave than any formulaically executed project plan. It will feel more like this:

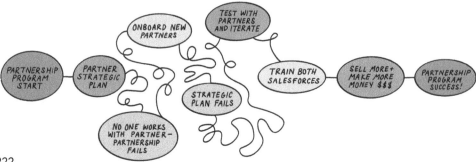

In fact, it might feel harder than your Launch Wave. Because during launch, you had one stakeholder you couldn't control: your customer. You had a small team, and they were all rowing your boat in the right direction and had the energy, dedication, and focus to ride that wave.

When you're building a partner program, you have three stakeholders you can't control: your customers (because in the end, you're still doing this for them), your partners (and their separate priorities and motivations), and your internal teams (that are now large in number and not all rowing in the same direction).

Doing Partnerships Well

Now that we've convinced you of both the value of building partnerships and the challenges involved, let's talk about how to do it well.

Approach a new partner program like you would a new product introduction, with the added complication of fitting a few other companies' ambitions and motives into the process. Here's a selection of the challenging steps you'll want to focus on initially to get your first partnerships off the ground.

Get the Right Team Together

This is a "huge and complex" project (recall our project types breakdown under **Don't Be a Grasshopper** in **Mistake 11: Managing Like a Startup Founder**). Identify a project leader who's going to be your partnership manager or director. This project leader must be a great combination of strategic outlook and operational project management, and this role must be their full-time job for at least a year (and hopefully longer if they enjoy the role).

Don't Try to Plan Everything in Advance

We've had experiences with teams that have designed entire partner levels (with benefits and associated fees) prepared website forms to "sign up to be a partner," and even purchased partner relationship management software, all before talking to their first potential partner — or even to their customers about what kind of partnerships would benefit them. Don't do this. It's a complete waste of time.

Define and Validate Your Hypotheses

What do you believe you'll accomplish from a successful partnership program? Here are some example hypotheses to get your creative juices flowing.

> ❯ We believe our customers will use more of [feature X] of our product, if they can launch it from [product Y], which they use every day.

> ❯ We believe our customers will buy more of our product if they could buy it through [partner X] and on [partner X]'s legal terms.

> ❯ We believe our customers will pay for post-sale support services from other service providers we certify.

Get your customer research moves on and reach out to customers to interview them and see if they validate your hypotheses. Don't waste your customers' time. Collect all your hypotheses together and get feedback on everything all at once.

A great way to get a full set of potential technology partners is to research your users' tech stacks to understand all the other solutions they use as part of their overall workflow alongside your product. (Ideally, someone in your company has already done this as part of defining your customer journey, discussed in **Mistake 12: Staying Within the Four Walls of Your Own Product.**) If done correctly, your interviews should uncover a set of common use cases to help you start thinking about requirements for the partnership.

Define Your North Star for Each Validated Hypothesis

You can't plan out every step of a huge, complex project like this. But you should define an ideal end state to help guide your team and align your organization. Your North Star will most likely change after you've learned much more about your needs, your customers' desires, and your partners' expectations. But that's not an excuse for not defining it now. Begin with the end in mind.

Reach Out to Potential Partners and Develop Personal Connections

Ideally, someone at your company has already been doing this as part of business development efforts. If not, don't waste any more time — start now. It's okay if you don't have a partner program pitch to give them. It's likely going to take weeks (perhaps months) to get any reasonable number of meetings on the calendar.

Your first meetings will be "first date" meetings — you'll tell each other about your products, share a bit about your business direction, and agree it makes sense to talk further. Then you'll exchange the mandatory NDA for signatures. Your goal in these initial meetings is to:

› Understand the potential partner's strategic direction and what their goals are for the next year or more.

› Confirm your hypotheses align with the partner's goals.

› Understand a bit about the partner's organizational structure.

› Determine the person with partnership authority.

> Your first meetings will be "first date" meetings — you'll tell each other about your products, share a bit about your business direction, and agree it makes sense to talk further.

225

Next, get out of your office and meet the key decision makers at your target partners. Find them at conferences. Invite them to your company's user conference (if you have one) or just to your office (if you have one). People do business with people they like. Get liked.

Get Your Internal Team Aligned and Pitch for Partnerships

It will feel like a herculean influencing effort to get all the internal teams we discussed in the prior section aligned. This isn't a discrete step. It's an effort you'll need to continue for the life of your partner program.

You are now ready to pitch the partnership to your target partners. Your pitch isn't an investor pitch, and it isn't a sales deck. It's a specific partner presentation that lays out the North Star of the partnership, the value each of you will receive from the partnership, the investment each of you will make into the partnership, a proposed timeline (which of course will change as you start implementing), and how you'll measure the success of the partnership. The end goal of this effort is a signed partnership agreement, and then off you go to implementation and onboarding.

Onboard Your New Partner

Properly onboard and train your new partner to ensure their teams (sales, marketing, perhaps product if an integration is being built) understand your company's position, your product offerings, your ideal users, and the combined partner positioning. Provide training materials, workshops, and support resources, and require the same in return from your partner. Foster a sense of commitment and loyalty between your two companies. There's no point in adding new partners if you can't make your first one or two successful.

Side Note: Quick Tips for Partnerships

You may have followed all the advice in this chapter, yet you remain unable to get a larger company to partner with you. Here's a hack we've used. Talk to people at your partner company to find out what their theme is for their next user conference. Find a way to position a combined workflow of your product with theirs in a way that promotes that theme. Pitch them the workflow as a presentation at their user conference.

Don't waste your time building partnerships with a company that has a substantially different go-to-market model from your business. If you sell a product through your direct sales channel to enterprise customers, and your target partner sells their product online to people who can just enter their credit card, it's unlikely that you're selling to the same user base. Partnership programs are supposed to make it easier for you to ride through the Scale Wave. If you can't leverage each other's customer base, the partnership isn't going to help you float — it might even be an anchor dragging you down.

And That's Why Partnerships Aren't Just a Side Hustle

If you're planning to scale your business through partnerships, then approach it as a new company initiative, not a budget expense for three more headcount. People from departments across your company will have to learn to exercise a different set of muscles. This is why it's almost always a bad idea to try to build partnerships as an early-stage tech startup.

If you haven't found your product-market fit yet, a partner isn't going to find it for you. If you've found your product-market fit (or business-market fit as we discussed in the Pivot Wave, **Mistake 7: Too Attached to Pivot**) but you haven't hit "the chasm" — selling beyond your early adopters to the early majority — you probably don't have enough scale to make any large partner pay attention to you. Don't expect to set meaningful revenue targets against partnership efforts this early on.

> If you haven't found your product-market fit
> yet, a partner isn't going to find it for you.

Now if you're a mid-stage startup with a material number of customers for your industry and an intention to invest in scaling to get large, then you're at the point you need to invest in a real partnership strategy. There are very few successful tech businesses that can build sticky products, sell across the globe, offer great post-sales service, grow their user base into other segments, and maintain their leadership without creating an ecosystem of partners.

Many startups approach building a new partner program the way people decide to exercise. They pay for the club membership and a trainer but only go for a few quarters before life (and the existing business) gets in the way. First, the trainer gets fired, and then the membership gets canceled.

You don't have to fix that leak in your boat with your own resources. The right partnerships can drive the growth of an ecosystem of other companies whose entire success depends on you selling more, which will make them want to help you do just that. You may be able to survive your first encounter with the Scale Wave without partnering, but you can't keep riding it alone forever.

Key Takeaways

1. Building a successful partnership program is like launching your second product. It takes the same amount of experimentation and upfront investment, and it will be roughly about the same length of time before you see the fruits of your labor. You'll hit the same waves: a Launch Wave, likely a Pivot Wave, and then a Scale Wave.

2. Technology partners help you scale through better customer retention (stickiness) by helping your product be where your customers live (be WOCL).

3. Building an ecosystem of smaller services partners that deliver great services to customers of your product will motivate these partners to encourage more companies to use your product, so they can expand their own service client list.

4. Partnering with a reputable reseller in a country you want to expand to can relieve you of the marketing, sales, and support you would have to learn to undertake going into that country alone.

5. To see a good ROI on your partnership efforts, you must leverage the investment you make to achieve one successful partnership (which likely isn't going to be your first one) into a playbook that will help you make many more partnerships successful.

Sidebar:
Partnerships Done Right

Atlassian started out as a product-led growth company with no direct sales team. They offered their software through their website and made it easy to purchase by entering a credit card online. They created a few packages and did a great job understanding their customer needs and building a product that met those needs. Their customers loved them, and a large part of their initial growth was through word of mouth.

They also generated communities early on, so their customers could talk to each other. That generated a lot more interest in their product, and they have a huge community today. It's a multi-billion-dollar company. Several years down the line, they've built a robust channel partner program, in which their partners sell the program.

It's easier to hire your own salespeople because you control them and train them. It's harder to be represented by a third party, and it requires you to do things differently. But it's a trade-off Atlassian was willing to make for longer-term growth. First Round Capital's podcast *In Depth* interviewed Jay Simons, former President of Atlassian, and recapped the following on their website:[10]

> *"We were focused on where we could get leverage that improves our speed and velocity and efficiency of the business. We didn't have our own sales team that was hunting for the best deals. Whenever there was a really big opportunity, we were motivated to introduce a channel partner of ours to handle the sales engagement," he says. It's now a core part of the company's marketing flywheel, and created a brand-new ecosystem, with over 400 Atlassian channel partners around the world that employ many thousands of people. "If it was just us, with a team of 100 people in Germany, that approach I believe is dwarfed by the thousands of partners in Germany that get up every day, without a business card from Atlassian, but are in essence working for Atlassian," says Simons."*

10 First Round, "Unpacking 5 of Atlassian's."

Scale Wave Wrap Up

Taming the Scale Wave means learning how to move from operating like a company that's hacking its way to a successful business model during launch or pivoting its business model within its current constraints, to a company that's efficiently exploiting its current business models.

As you scale, so will the wave you're riding — at least it will feel that way. It'll feel like multiple waves hitting you, quarter after quarter, and even year after year, depending on how long it takes you to reach calmer seas.

Scaling is hard work. If it weren't, you'd see more startups breaking through the Scale Wave, emerging as multi-hundred-million annual revenue companies and finding their way to calmer waters or the last of our waves — an exit. In our experience, the work is 90 percent excellent execution and only 10 percent strategy. And execution mistakes like the ones we've set out here won't be solved by brainstorming your way to a new strategy.

Navigating the Scale Wave will feel scary — scarier than during your Launch Wave — because now you have many employees, customers, and perhaps investors relying on the company you've helped build.

But when you succeed in sailing through the Scale Wave, magical things happen:

> The chaos dies down. You'll have helped your teams invest in systems, processes, and perhaps different people. This allowed you to get out of firefighting mode.

> Your velocity increases. You've re-architected your product, your team, and your approach to customers. Now you're set for your 10 times challenge, instead of your 10 percent KPI.

> Your financial metrics improve. Your internal efficiency improves, and your external programs with your customers and partners have finally allowed you to realize economies of scale.

The Scale Wave is daunting, and it will try to beat you down. But you're not the type to get seasick. And you don't give up. You might make some mistakes along the way, and that could cause you to take on water and slow your progress. But if you avoid the mistakes we've described, you'll stay on course, and the right team will stay on board with you.

We close this chapter with one of our favorite quotes, from Charlie Munger, who was the vice-chairman of Berkshire Hathaway.[11]

"It is remarkable how much long-term advantage we have gotten by trying to be consistently not stupid, instead of trying to be very intelligent."

---

11 Chew, "Charlie Munger," Constant Renewal.

The Rogue Wave

The Exit

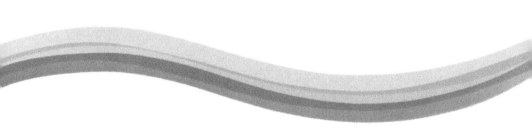

Entrepreneurs and intrapreneurs are in the business of innovation and growth. We bring new things to life, we persist through pivots, and — if we're lucky — we learn how to take our big ideas from concepts to launched businesses and through to scaled operations. We conquer the waves that hit us at each stage because we're warriors.

But very few of us are corporate transactional experts, and this fourth wave, the Exit Wave, is an unusual challenge — one that's usually shrouded in inscrutable legal, financial, and tax details that don't appear to have anything to do with building great businesses.

When the business community talks about an exit, they're referring to an initial public offering (IPO) or an acquisition. In both cases, the owners of the business are usually (there's always fine print to consider) able to convert their ownership from paper they can't sell or trade, to a "liquid" asset they can turn into cash. The business world is obsessed with IPOs. And if you can take your company public, you're in rarified air, achieving what very few businesses manage to achieve. In fact, only 3 percent of seed-funded tech startups in the United States make it to a size that allows them to go public. On average, two-thirds of them appear to either go out of business or hover in private purgatory, unable to pay back their investors or make good on the options they granted to employees. The rest? They find an exit by getting acquired, sometimes successfully, sometimes less so.

Silicon Valley venture capitalists don't like it when their startups talk about preparing to be acquired. They tell their portfolio of startups to swing for the fence and aim to be IPO-ready — and this ethos permeates into the minds of entrepreneurs, driving many to aim for building only a large, enduring, stand-alone, public-company-worthy business. What could go wrong with that plan? The story is in the numbers.

The same people who tell you not to spend time planning to be acquired will also tell you that if you build a successful company, you'll have many exit options. This guidance sounds a lot like, "If you build it, they will come." While this worked out for the character Shoeless Joe Jackson in *Field of Dreams* (1989), that strategy's best left for the movies. Anyone who has achieved anything difficult in life knows things rarely just happen when you put your head down and work hard. If you want to build a company with many exit options, you should set a strategy for that, and then operate with that strategy in mind.

> If you want to build a company with many exit options, you should set a strategy for that, and then operate with that strategy in mind.

What if a large swath of these two-thirds of seed-funded tech startups that are currently falling into private purgatory or closing up shop were instead advised to focus on building a company that larger companies wanted to acquire? What if non-tech companies did the same? We believe if every company developed its acquisition options the way it developed its business as a standalone company, then we'd see more startups having successful exits through acquisitions instead of being stuck in private purgatory or going out of business. They would bring their products to a broader market; give employees increased options (by either through realizing the value of their equity on exit or continuing with the acquiring company in secure employment); and reward more founders for their hard work in navigating their startup ship through the three formidable waves of Launch, Pivot, and Scale.

> If every company developed its acquisition options the way it developed its business as a standalone company, then more startups would have successful exits through acquisitions instead of being stuck in private purgatory or going out of business.

If you're committed to aiming for an IPO, you'll likely be riding the Scale Wave for quite some time. You'll move beyond the scope of *Sail to Scale* and need to start developing your team's expertise in "IPO readiness." While an IPO is an exit for your shareholders, it's not really an exit for you and your team. You'll continue to run a company that will need to continue scaling. If, however, you aren't willing to bet your company on the incredibly low odds of going public, we're here to tell you that's okay. You can still build a successful company that can deliver value to the larger ecosystem by getting acquired. And if your investors don't want to hear you talk about your acquisition strategy, well . . . don't talk to them about it. That doesn't mean you shouldn't prepare for it.

To get there, founders and leaders must face this last of our Four Waves. It's perhaps the most imposing because it's one they rarely face, and its timing is not entirely within their control. We call this last wave your Rogue Wave — a rare ocean phenomenon that's twice the size of the waves around it. Rogue Waves, like acquisitions, are difficult to predict and can be dangerous to your corporate ship if not approached with skill and preparation.

In this last section, we review four big mistakes startups make as they consider an exit through acquisition — mistakes that, if made, will have them staring down the precipice of a rogue wave.

In **Mistake 15: Running Out of Runway**, we explain why M&A is a strategy, not a transaction. **Mistake 16: You're Inside-Out** and **Mistake 17: Selling the Future of Your Business Instead of Theirs** show how to go about developing that strategy, as well as the two biggest mistakes startup founders make when building it. Lastly, in Mistake 18, we teach you how to think like an entrepreneur, not an investor. **Mistake 18: Rejecting the Right Acquisition Offer** also includes two case studies and cautionary tales of what happens when you ignore the signals potential buyers are sending you.

Story:
Building for a Good Goodbye

Ashley McLain met her co-founder, Larry Cox, at an environmental consulting firm where they both worked. After a decade there, Larry approached Ashley with an observation. They had all the leadership-level responsibilities but none of the associated authority. They didn't have corporate budgetary authority, and they weren't part of the strategic and financial management of the business. They approached the owners to test whether there was room for them to be more invested in the company. It quickly became clear the owners weren't interested in succession planning of any kind.

Ashley and Larry resigned on a Friday in 2007; they spent the weekend with their third partner, Lorie Cox, putting together Ikea furniture in a small office; and started work as Cox I McLain Environmental Consulting, Inc. (CMEC) the following Monday. The Crack in the Market they saw was a set of customers that was limited in who they could engage with due to the procurement requirements they were obligated to comply with. As two of the three founding shareholders were women, they qualified as a women-owned business and spent the time to get certified as one. This qualified them to compete for a variety of projects involving public money in which agencies were required by regulation to award a certain percentage of work to certified companies owned by underrepresented groups.

Like the Newsela founding team we discussed in **Mistake 10: Staying in a Constant State of Firefighting**, CMEC understood growth hacking. Ashley recalled that they had one hundred meetings with potential clients in the first six months of their new company, sometimes driving four hours each way to attend one meeting.

"We were yin and yang," Ashley reminisced. "Larry was rural; I was urban. Larry's background was ecology; mine was the human environment. At conferences, Larry would have two 60-minute conversations. I would have 60 two-minute conversations."

As they worked to build their client pipeline, they also knew they had to assemble the right team with specific expertise that would differentiate them from other environmental consulting firms. They were thoughtful about hiring. They first added cultural resources (archaeologists), which rounded out the core categories with socioeconomics and natural resources. Gradually they added architectural historians, endangered species experts, Geographic Information Systems mappers, geologists, wetland specialists, and planners. They hired from a variety of universities known for different environmental specialties. They were careful to expand methodically into growing cities with few certified firms, where they had established the trust of engineering team partners. Together, they could win competitive pursuits by fulfilling required certification percentages through the qualifications of their team.

The business was growing, and the team was working well together. A decade after founding the company, a mentor of Ashley's who ran a successful engineering firm told her it was time to hire an outside consultant firm. She saw that CMEC was facing the Scale Wave and didn't have the experience to navigate it themselves. Ashley and her partners took her mentor's advice. The consulting firm they hired gave CMEC some advice that was worth the fees they charged: *If they ever wanted to be set up for a successful acquisition, then they needed to develop the strength of their second level of leadership, since an acquirer would assume the founders wouldn't stick around for years and would want to know there would be leadership continuity.*

The founders recruited three rising stars, each with specific skill sets — archeology, biology, and history — and assigned one each to marketing and business development, learning and growth, and operations and performance metrics. The common denominator was their entrepreneurial mindset. With some advice from an outside financial consultant, they provided bonused ownership interests in the company to these three individuals and, as Ashley put it, "We did our best to 'brain dump' our knowledge onto them." Through all of this, the founders made sure they held a majority of shares in the company, so they could efficiently control how the company chose to exit.

CMEC also started getting intentional about investing more in the business of managing the company. In contrast to some small firms, where employees ride the coattails of the founder's reputation, CMEC focused on what Ashley called "building a grown-up business." They invested in getting out of firefighting mode (see **Mistake 10: Staying in a Constant State of Firefighting**) by hiring additional finance experts, implementing productivity software, and spending more time communicating and setting expectations with the team. It would've been easier not to invest in these scaling efforts and focus on growing the business by getting more projects and clients. But the founders were taking the advice of their consultants seriously.

They invested in getting out of firefighting mode by hiring additional finance experts, implementing productivity software, and spending more time communicating and setting expectations with the team.

The firm was growing large, with nearly one hundred full-time and contract employees across five different offices. Things were going well. Companies were flocking to Texas for the business-friendly environment with no state sales tax, so infrastructure was growing. And then . . . COVID-19 hit. Unlike other companies, the pandemic didn't hit their revenues. Knowing they couldn't afford to stop conducting fieldwork, they pivoted to people driving alone in their cars, packing lunch, grabbing whatever hand sanitizer was being produced by the local distilleries, and hitting the field. Without fieldwork, they would have no regulatory compliance reports to write. Without environmental clearances, engineers couldn't complete design projects, and funding couldn't be released for construction.

When President Biden's infrastructure bill passed in 2021, it created an additional tailwind for CMEC, as the bill spurred even more large infrastructure projects that required environmental regulatory compliance. Projects kept coming in because CMEC experts knew how to complete legally defensible reports under streamlined expectations, which had become federal policy.

It was about this time Ashley, Larry, and Lorie started seriously thinking about looking for an acquirer. They'd been at this now for fourteen years. They understood that to steer the CMEC ship through the next part of its journey, they would need to expand geographically, strengthen their financial capabilities, become more sophisticated in HR compensation, improve performance management, handle more employee relations issues, and invest in more professional development and training.

Ashley loved her company, the people who worked there, and the culture they'd developed. "But I really missed the fundamentals of environmental consulting — the projects, problem-solving for clients, and working with our people," she said.

 Ashley and Larry felt they didn't want to be the "ceiling" on staff members' or the company's ability to continue growing. They knew much of the scaffolding they needed to build out would already exist at a larger company.

In spring 2021, they decided to hire an M&A advisory firm to help them think about how to approach selling the company. Between President Biden's infrastructure bill and financial institutions' newfound obsession with ESG (environmental, social, and governance) profiles of project developers, suddenly environmental consulting became "sexy" — well, as sexy as that industry could get. The work they'd been refining for such a long time suddenly became increasingly critical for a growing number of architecture and engineering firms to build new communications infrastructure, transportation, and utilities projects. It was a good time to look for an exit.

They interviewed three M&A firms. Ashley recalled advising her partners that the fee the firms were charging was absolutely worth it if the business could achieve a successful exit. Sometimes saving money can get in the way of making money.

The company hired one of the three advisory firms based on their deep understanding of the field of potential acquirers. The firm helped the founders translate how they talked about their work and their business into what Ashley called "market speak." The consultants helped them develop corporate presentations that spoke to the KPIs — key performance indicators — the field of acquirers would want to know, rather than what the founders might want to talk about. Ashley shared:

> "Looking back, 2021 was a whirlwind of confidential communications with a list of nearly three hundred potential acquiring companies, over sixty receiving the confidential profile, more than forty responding, seventeen signing legal NDAs for data room access, a dozen interviews/in-person meetings, and an overwhelming number of diligence conversations."

In the process, Ashley realized the large firms she and Larry knew best weren't the ones that ended up being the most interested in them. They knew firms in Texas, where their headquarters were. The acquisitive firms that came to the table to negotiate with them were located in other states and were looking to expand out of their regions into states like Texas with large opportunities for growth. It was their M&A advisors who helped them identify and connect with these other firms.

In the end, CMEC received five acquisition offers. Ashley chalked up this wildly successful result to the hard strategic work they did building their firm:

> "Yes, we had a great reputation and built a strong culture that kept people there. But we were very intentional in building a business that we and, in turn, an acquirer would value. Acquirers saw that we could attract bright talent. We had built up specialties in the environmental sciences that many firms didn't invest in. We had key customer contracts that were relevant to large firms and an entrepreneurial second level of leadership that would likely stay after the acquisition closed."

Ashley and her partners had prepared for an acquisition long before they even knew they wanted to sell. They also figured out that the innovation and agility people need to succeed at a small company could give an infusion of energy to a big company.

The founders didn't pick the highest offer — a pure private equity bid that felt to them like a less-than-perfect home for their employees. With guidance from their advisors, they picked the offer that was a strategic fit, that would clearly benefit from the value CMEC could bring, and that didn't lock the founders in for years after the acquisition. Their five geographic locations enabled the acquiring company to achieve a more complete geographic coverage in the US, while allowing CMEC staff the chance to work on bigger projects in a wider geography at a top ten international design firm. And CMEC's technical expertise and client relationships were additive to the acquirer's services.

But the story never ends at the offers.

The next phase of the acquisition process involved deep due diligence. Ashley recalled this being one of the hardest things she had to go through:

"We called the entire diligence process a 'corporate colonoscopy.' And one of the hardest things for me was the secrecy of it all. We couldn't tell the rest of the company what was happening in case the deal fell through, which in a couple of instances it could have. So I had to live this secret life of spending weekends uploading documents and answering diligence questions, and weekdays handling clients and attending virtual socials to keep the team together during COVID. I have trouble keeping birthday presents a secret! Thank goodness I had a staunch commitment to closing and a good therapist!"

The machinations that happened to complicate this particular acquisition could take another ten pages to explore. The truth is that every acquisition is a complex, confrontation-ridden, almost near-death experience. But if you're determined, persistent, practical, resilient, and lucky, the deal eventually closes. Ashley's deal did.

Mistake 15: Running Out of Runway

CMEC's story is a case study for how to build your company while creating the perfect setup to maximize your options to exit through acquisition. If you only read business headlines, you might assume all companies who try to get acquired manage to do so successfully. You'd be wrong. We'll compare CMEC's growth approach to that of other startups that didn't manage a successful acquisition throughout this section. This first mistake we'll cover is one of timing rather than approach.

Mona had experiences with startups that waited too long during her time as an M&A advisor at Tribal Ventures. Mona often started conversations with startup founders by bringing up Y Combinator co-founder Paul Graham's iconic startup curve circa 2008. The image's witty insight reflects Paul's drumbeat of advice that if building a company doesn't involve a certain degree of pain, then you probably aren't trying hard enough. The big question is whether you can ride out the pain (ride through the waves) and make it through to happiness. It's not just a matter of hard work. It's a matter of timing.

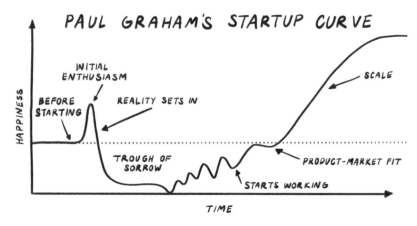

Source: Paul Graham

At Tribal Ventures, the team modified Paul's startup curve into a Startup Valuation Curve, to reflect that valuations for startups don't continuously go up and to the right as their founders would like to believe. Like the level of founder happiness, the value of a company ebbs and flows as the company hits new waves, takes on water and risks capsizing, and, with skill and perseverance, masters the wave and finds itself at a new high point. Valuation high points in your startup's journey are often associated with two things: the level of your happiness and excitement about your startup's opportunities and your startup's likelihood of negotiating a successful acquisition.

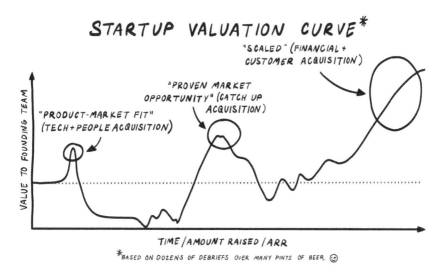

Source: © Tribal Ventures, LLC

Unfortunately, most founders Mona's team met waited until they hit their troughs rather than their peaks, and then, feeling pain and desperation, sought an exit right at the time their startup risk was highest.

Desperate Times Call for Lower Valuations

Successful acquisitions are typically the result of years of preparation. Yet most startups, whether due to lack of advice or bad advice from advisors and investors, don't think about their acquisition options until they can see the end of their cash reserves and realize their chances of raising more money are low.

Successful acquisitions are typically the result of years of preparation.

It turns out your odds of raising money are highly correlated to your odds of negotiating a successful acquisition.

Let's look at the first peak in the Startup Valuation Curve in more detail. Investors are looking for a reason they should believe you're much more likely to succeed than other companies they could invest in. In your Launch Wave, you give them a reason to believe by showing them you've found that elusive product-market fit. Showing product-market fit de-risks your idea, your product, and your founding team. If you're lucky enough to raise money before you can show product-market fit, it's likely because someone believes in you and your ability to figure out how to build a business. Most of the time, founders who try to raise money before they achieve product-market fit are met with frustrating comments from investors like, "We'd like to see you get more traction before we invest," or, "Once you've built out your technology further, give us a call." This is code for, "We don't believe yet."

Showing product-market fit de-risks your idea, your product, and your founding team.

Acquirers are also looking for reasons to believe — especially when looking at early-stage companies navigating their Launch and Pivot Waves. Certainly, there are stories of companies that have been acquired before they reached product-market fit — recall Twitter's (now X) acquisition of Vine, a social networking platform for sharing six-second video clips, a few months after it was founded in 2012 for an estimated $30 million. But these are outlier scenarios that shouldn't form the basis of a founder's strategy. An acquirer buys a company in its Launch Wave when the acquirer is looking for specific technology it would prefer not to build in-house. It also might acquire when it needs a small team with a specific set of expertise, and it wants to de-risk hiring a team that can't work together or won't get the project done. Whether an acquirer realizes it or not (and oftentimes, acquirers justify their acquisition decisions without really understanding their underlying motivations), startups that are able to show an acquirer they have product-market fit help the acquirer believe they're buying a successful product and team.

When investors and acquirers believe, they tend to be ready to pay (invest or buy). This peak of belief for investors and acquirers aligns with the peak of happiness and confidence for founders. At the same time an acquirer might be interested in buying a startup, the startup's founders are confident they can go on without having to sell. That's what pushes the valuation of the startup during this period to its peak.

> This peak of belief for investors and acquirers aligns with the peak of happiness and confidence for founders.

Whether a formal acquisition offer is made at a valuation peak, or whether there's just a possibility a startup can get acquired even if the founders aren't actively looking, the end result is often the same. Founders, all happy and confident, forge on, bypassing (possibly neglecting) an acquisition option in favor of raising money to build their large, enduring, stand-alone public-company-worthy business.

Regardless of which valuation peak a company is at, the next part of the story is often the same. As the company moves from having conquered one wave to approaching the next, leadership must learn new skills (or be replaced), new hires must integrate with the Old Guard, employees must approach running the business in different ways from how they were operating through the last wave. People are generally bad at dealing with change. Employee engagement sinks. The company hits the "trough of sorrow."

Whether the company is learning to do the right things to scale the next wave or not, the pain is the same. Building a business is hard. If it wasn't, then many, many more would succeed. The pain is magnified if the company didn't raise enough in its last round to give it the room needed to survive this trough. Not only is the work painful, but the company also might run out of money before it can hit the next milestone that supports the next funding round. It's during this trough that leaders will suddenly come up with a brilliant option for their business. "Hey! Why don't we try to get acquired?" certainly sounds better than continuing to suffer through the feeling of impending doom.

Wanting to Sell High, but Having to Sell Low

"Buy low; sell high" is possibly the most famous adage about making money in the stock market. It's so obvious that it sounds like a joke. It's a lot easier said than done.

When companies look to get acquired because they're running out of runway (whether the runway is their energy to continue or the number of months left until they run out of cash), they're effectively looking to sell low — the opposite of what the famous stock market adage tells you you're supposed to do. And when we say "low," we mean even lower than your theoretical valuation a year ago when you hit your last peak and felt like you had all the options available to you. This can be difficult for a founder to understand.

It doesn't seem right that your valuation should drop even as you've probably brought on more customers and built more features. And yet it does. The risk that someone valuing your company thought you had mitigated when you had proven product-market fit is now replaced with a new risk — that you haven't figured out how to build a financially sustainable way to acquire your customers. This new risk hits you after you've cleared your Launch Wave, perhaps as you're coming to realize you're hitting the Pivot Wave, or as you're first preparing for the oncoming Scale Wave.

Regardless of timing, when the next risk does hit, acquirers and investors doing due diligence on your company will see it. You'll know this is the risk they're focusing on when they start asking you about your customer acquisition cost (CAC) and other key metrics (go to the end of Mistake 1 to read the **Sidebar: SaaS Metrics**). If you're spending more getting your customers than the value of your customer contracts, an acquirer isn't likely to take the chance that they'll be able to crack the magical go-to-market motion you haven't managed to figure out. At best, they'll hang back and wait. (At worst, they'll decide they aren't interested — now or later.) Once one potential buyer steps away, others will follow — not necessarily because they should, but because acquirers are run by humans who have biases. Conformity bias — the tendency to act based on the actions of others rather than one's own independent judgment — will drive a potential acquirer to think less of your business if they learn that another company wasn't interested.

This new risk you haven't managed to show you can overcome combined with your short runway, along with one or two market participants possibly seeing the risk and the short runway — and, well, running away — all compound to bring down your valuation just as you're finally thinking about selling your company.

Selling as You Run Out of Runway

When Mona worked as an M&A advisor at Tribal Ventures, she worked with a few startups that had decided they needed to sell their business because they were running out of the cash and energy needed to continue as a stand-alone company.

The founders would come to Mona with high hopes and even higher valuation expectations. When Mona's firm did their own diligence, they'd discover these companies had less than nine months left of runway. Nine months is a very short period of time for start to acquisition close. And a rushed acquisition process generally leads to a much lower valuation than the founders have in mind.

Nine Months to the End of Your Runway

The founders would insist they had already been talking to potential acquirers and believed those acquirers were interested. Mona had been the head of corporate development at multiple companies. Here's what she knew and what the founders she was working with didn't understand: It had been her job at those companies to take calls from founders and hear their pitches. It's what she was paid to do. But that didn't mean she was interested in their company.

Many founders will mistake a conversation as interest from a company. Sometimes, that's the case. Many times, the company is just exploring what's out there and assessing the landscape. Rarely (but still sometimes) the company might be using this conversation as an opportunity to validate their interest in another acquisition, or to collect more information about the competition. And even if they are interested, they're always juggling multiple priorities, meaning that chances are low they'll stop everything to make this acquisition their new priority. Every email exchange takes a week to get a response. Every meeting takes two or more weeks to get scheduled. And time is slowly running out.

Founders with a short runway end up spending more time than they anticipated reaching out to potential acquirers and creating pitch decks. They're reaching out to investors to try and raise money to extend their runway. If they don't have good advisors, they're using the same pitch deck for both types of conversations. The pitch deck for investors speaks to how much faster their startup would grow with an additional investment. But let's remember, any corporate development type at a big company is not in the business of investing in startups. They're in the business of acquiring startups that would allow them to better invest in their own company.

Investors want to know how much faster your business will grow with additional investment. Acquirers want to know how much faster their business will grow if they acquire you.

The pitch deck for an acquirer must focus on how the startup can help the acquirer grow its own business. In the same way that startups pitch better the more they pitch to investors, they also improve the more opportunities they have to pitch to acquirers.

If they have good M&A advice, the startup has also put together pro forma financial projections that show how their business can add value to the acquirer's, and they're sending reams of follow-up emails trying to get more meetings with their target acquirers. They probably haven't had any time to put together a data room yet.

Six Months to the End of Your Runway

All the work we just described takes about three months to get through. Around this time, a potential acquirer would have had enough internal conversations to make a decision about whether to take the next step with this startup or not. If not, they'll send a polite email letting them know this wasn't the right time to enter into a discussion. If they're interested, then they'd continue the conversation by asking for a set of "pre-diligence" information (see the **Sidebar: The Pre-Diligence List** after Mistake 17). What this requested information includes depends on the reason for their interest.

A company Mona worked at was interested in buying the technology of a startup as a tech plug-in, so they didn't have to build the solution themselves. The startup was very small and was selling to individual users at similarly small companies and other organizations. Mona's company was selling to enterprise customers. The startup's customer list didn't present much value. What she cared most about was avoiding upsetting the startup's customers as her company moved the technology onto a more expensive platform. Other than that, her pre-diligence questions were all about the chain of title to the technology itself. Who built it? What agreements did they sign? Where were these people now?

The focus was different with another potential acquisition. The startup's product was more than just a "plug-in technology." It was needed to help Mona's company compete with other key competitors, so she cared very much about the startup's customers as well as its product. She cared about whether the customers were renewing their use of the product — because long-term customers showed that the product was competitive. This startup was running out of runway, and her corporate development team had only a few weeks to run through a great deal of anonymized customer data to decide if they wanted to take the next step and deliver a term sheet. The team spent a month doing detailed analysis that led them to determine that most of the startup's customers were churning. They were just very good at getting new customers to keep their revenues flat, even as existing customers left them. Mona and her crew bowed out of the opportunity.

If you're lucky enough to get to the pre-diligence stage of discussions, it'll go on for a month or perhaps two. But in our timeline, you now have only about four months left before your ship runs out of steam and starts to sink. If you don't have serious buyers with four months of runway left, and you can't raise money to continue funding your operations, you're likely looking at a fire sale — meaning you're selling your assets, and perhaps packaging up your team as an "acquihire," at a distressed sale price.

An acquihire is a transaction in which a company is bought primarily to hire the employees working there as a group. Acquihires often also include an acquisition of technology or product, but only as a secondary motive.

The key to understanding the enormity of the impact of running out of runway is understanding the influence of a triggering event. Your startup has a triggering event that makes you want to move quickly — you're running out of cash. Potential acquiring companies likely have no triggering events — they're not facing anything that forces them to move quickly to buy you, except maybe if they really wanted to hire your employees before they jump ship and take other jobs. When you have a triggering event but your acquirer doesn't, your acquirer has all the leverage.

 The person with the triggering event has no leverage.

Mona didn't much enjoy negotiating founders down to such a low acquisition price. She had been a founder once and understood how much work goes into building a company, even when that company eventually fails to succeed. But the job of an acquirer is to get the best price they can. And at this somewhat desperate point in a startup's journey, there's often limited value a larger company can extract from the startup's products, team, or customers. In one instance, she recalled her company had refused a price proposal from a selling startup three months earlier, only to have them come back around later suggesting half the original price. She ended up closing the transaction one month before the startup ran out of money at 20 percent of the originally proposed price.

The Last Three Months

With just one quarter left of runway, you're at the end of the road. If you have a company lined up to buy you, then it'll still take a couple of weeks to negotiate and sign a term sheet, a couple more weeks for the acquiring company to do at least the most basic of due diligence, and a few more weeks for the lawyers to draft up the paperwork, negotiate it, and prepare to close the deal and transfer funds. Work expands to fill the available time. So you'll inevitably find yourself getting disturbingly close to your last days of operation while constantly emailing lawyers and executives at the acquiring company, pushing them to work faster.

Compare this scenario to the story of CMEC, the environmental consulting firm with multiple options and no triggering event, and you'll understand why it's important to build an M&A strategy long before you need it.

M&A Is a Strategy, not a Transaction

The majority of startups that get at least some funding still fail to either keep growing or find an exit. They inevitably close up shop or continue in private purgatory. It doesn't have to be that way. If you've made it through any of the Four Waves, you have a product that at least some customers love, and you've shown that your team has the resilience and skill to innovate and adapt, then you should be able to sell this business — and deliver a team to help grow it — to a larger company.

We believe the reason so many businesses fail to exit is because founders, often in reliance on bad advice, view M&A as a transaction that happens when you're ready to sell, rather than as a business strategy. This mindset drives too many startups into the doom spiral of decreasing valuation we've just described. M&A is a strategy, not a transaction.

In the next chapters, we'll cover mistakes people make in developing this strategy and how to avoid them.

Key Takeaways

1	Your odds of raising money are highly correlated to your odds of negotiating a successful acquisition.
2	It doesn't seem right that your valuation should drop even as you've probably brought on more customers and built more features. And yet it does. The risk someone valuing your company thought you'd mitigated when you'd proven product-market fit is now replaced with a new risk — that you haven't figured out how to build a financially sustainable way to acquire customers.
3	A pitch deck for investors speaks to how much faster the startup would grow with an additional investment. The pitch deck for an acquirer must focus on how the startup can help the acquirer grow its own business.
4	The person with the triggering event has no leverage.

Sidebar:
Should you hire an M&A Advisor?

A former M&A advisor herself, Mona still doesn't recommend hiring M&A advisors or investment bankers for every acquisition transaction. As with most advisors, you must first understand what gaps you're trying to fill before you go off in search of more people to add to your deal team. And like in most industries, there are good M&A advisors and not so good ones.

There are probably four scenarios in which a startup should consider filling a gap by hiring an M&A advisor to drive a successful acquisition of your business.

1. You're pressed for time.

This happens when you haven't treated M&A as a strategy, but rather as a transaction. And for various reasons — perhaps you're running out of runway, or perhaps you're receiving inbound interest from a potential acquirer — you need to move quickly. Recall Ashley's "whirlwind" sale of CMEC. Ashley credited their M&A advisor for helping them manage the overwhelming diligence conversations and much more.

If, on the other hand, you aren't looking at multiple potential offers, and you're not under a timeline, then you might have the bandwidth to manage both your acquisition strategy and process with your internal team. Whether that's the case depends on whether you can fill the following gaps.

2. You already have inbound interest and want to generate a bidding competition.

If you're one of the lucky startups that's approached by a company looking to acquire you, it could be the right time to hire an M&A advisor. Be sure that person is familiar with your market segment, because one of the primary jobs of M&A advisors is to have built relationships within acquiring companies, so they can move fast and generate other offers if a startup is looking for competing bids.

We've already discussed conformity bias (the tendency to act based on the actions of others rather than your own independent judgment). Conformity bias works against you when people know a company wasn't interested in acquiring you. But it also works for you when people find out a company is very interested in acquiring you. Strong M&A advisors are good at leveraging conformity bias by letting other potential acquirers know there's someone else at the table, so they need to move quickly if they're interested.

3. You don't know who the best acquirers might be.

If you haven't done your homework, and you don't know which companies to approach to propose an acquisition, an M&A advisor focused on your industry segment is probably your best bet for filling that gap. Within an M&A advisory firm, there will be people whose primary job is to develop relationships with all the corporate development people, the heads of product, and the CEOs of companies in certain market segments.

In CMEC's case, they chose their M&A advisor partly because of the advisor's national expertise in CMEC's market segment. CMEC thought their acquirer would be a company within their state. It was their advisors who helped them realize they were much more valuable to companies in other states looking to expand.

4. You aren't the best negotiator.

This last gap could be the most important one to fill, and it's often the one that's most frequently underestimated. We've never heard a CEO say, "Honestly, I'm not a great negotiator." And the qualities that often make a great CEO are the very characteristics that make for a very bad negotiator. In the Launch Wave, we explained the need for startup founders to be futurists — to imagine the future. We also identified a common mistake of startup founders — being in love with their solution. These kinds of startup CEOs (and even scaling CEOs) generate excitement in others and lead teams to explore and problem solve. They tend to be people pleasers — they like being everyone's friend. They also tend to be passionate. They get very attached to ideas, especially their own.

None of these characteristics have been connected with great negotiators. Negotiation requires a kind of dispassion and toughness that tends to be in opposition to someone with the innate capability to lead and inspire people.

If you don't have a strong negotiator on your team (and you aren't likely to be aware of it if you don't), a good M&A advisor should fill that gap for you. Experienced M&A professionals know when to push, and they know when to back off. They know how to phrase things so the conversation can continue while you're essentially saying no to a key proposal. They know how to help you understand what your walk-away-point is, and then remind you of it when you're hovering close to that point.

In addition, M&A advisors can take over the difficult conversations with your acquirer, leaving you to maintain a positive relationship with your soon-to-be new employer.

How to Choose Your M&A Advisors

Just like dentists, financial advisors, and lawyers, there are lots of average to bad M&A advisors. And just like these other professions, you should ask for references and interview M&A advisors before choosing one.

A good M&A advisor will meet with you, understand your business model and where you are, and ask you a lot of piercing questions about your customers, financials, and product differentiation. They'll be deeply curious about your business. Becoming an advisor to your company should be a long-term (probably a year, perhaps more) commitment, and a good advisor should be careful about where they invest their time. If an advisor is more than happy to sign you up without doing their own diligence, that should be a red flag for you.

A great M&A advisor has contacts in your market segment and comes to your first meeting with an understanding of the mindset of some of your potential acquirers. They know this because they're in regular communication with these companies and have developed an understanding of how they think. In your initial interviews with potential advisors, dig into who they know in the industry and what they know about the mindsets of those people.

The best advisors for your business are firms that will help you with the operational work of getting to a deal. Many of the larger M&A firms will help you with your deck, create financial projections, and facilitate introductions. But when it comes to helping you present the details of your company through the diligence process or being by your side to explain the operational ramifications of certain deal terms, they aren't anywhere to be seen. The hard work doesn't end when you get a term sheet. In fact, the process from term sheet to closing a transaction is often described by founders as one of the most harrowing experiences of their startup journey — like being hit by a Rogue Wave.

Recall Ashley's description of her diligence experience — it was a "corporate colonoscopy" shrouded in secrecy and conducted on weekends while the weekdays were business as usual.

If your advisor can't take some of that stress off you, so the stress doesn't destroy the deal, then they're a high-priced matchmaker. That still might be worth their fees if the end result is a successful one for your company, but it won't make the process any more manageable for you.

Mistake 16: You're Inside-Out

In the previous chapter, we explained why M&A is a strategy, not a transaction. The next two chapters show how to go about developing that strategy, as well as the two biggest mistakes startup founders make when building it.

The first of these mistakes is what we call the inside-out model. (Or, when we're with our tech friends, we may refer to it as "The Sims Effect" — creating an entire world filled with strategies from your own mind without any real-world input.)

Inside-out is a mental model that nearly always fails. It fails if your business is facing a Launch Wave and you're in love with your solution instead of focusing on your customers' pain points. And it fails when you're developing an M&A strategy by brainstorming with your internal team all the reasons Disney or Microsoft or Samsung would love to get access to your business.

While this may sound like something you'd never do, almost all companies do it. They've spent so long focused on building a sustainable business around their own product and strategy, they become very attached to it and so involved in the details of their world that the muscle to think about anyone else's has atrophied. It's a similar syndrome to the not invented here (NIH) syndrome, in which companies (usually general managers of product divisions and heads of engineering) are so wrapped up in their own work they can't believe anyone else could build a product better than them.

The inside-out model can lead you to mistake politeness for interest. A founder shows up with slide decks with nice graphics and compelling charts showing the perfect synergies that could be achieved if Company A were to buy their startup — or how compelling it would be for at least five large companies to acquire the startup's business. Perhaps the startup has even pitched this to one or all of these large possible acquirers. And the acquirer has expressed some degree of interest, likely misinterpreted by the startup as enthusiasm. We call this "happy ears" because we hear what we want to hear.

But as discussed earlier, it's the job of the executives at these acquisitive companies to talk to the startups in their market segment, so they keep up with the landscape of innovation and growth. And it's not the job of these executives to destroy relationships by shattering dreams. Like investors, they'll nod vaguely, and tell you they'd like to see more traction and you should keep in touch.

If you've had a conversation that sounds like this, where nothing has happened as a follow up, you were likely in an inside-out mindset. That's okay. Don't beat yourself up over it — we all do it. Read on to help you move out of that mindset and start thinking outside-in.

If you think back to the story of Hitch and their pivot to skills intelligence, Heather was operating with an outside-in mindset. Heather started her CEO role at Hitch not by getting her leadership team in a multi-day offsite to build a new strategic plan, but rather by researching and listening to customers. First she looked at what Hitch's peers in the market offered and how their customers were actually using Hitch's product. Then she spent two months on a listening tour, talking to as many customers and industry participants as she could, asking open-ended questions. All this outside-in exploration was what led Heather and her team to realize that companies needed the Hitch product for a different use case than what they'd been pitching. And those companies weren't just potential customers. One of them ended up being an acquirer. Hitch's skills intelligence technology was a product gap ServiceNow was looking to fill, as Heather had predicted by looking at the market from the outside-in approach. ServiceNow acquired

Hitch in 2022, an amazing outcome from a pivot that Hitch had conceived and executed only a year earlier.

Becoming Outside-In

You wouldn't try selling a product without first understanding why customers would want to buy it. Yet in our experience, startup founders regularly fail to understand what drives a company to buy another company. Like customers, acquirers are more likely to buy a company that solves their pain point rather than one that helps them be a bit better. So as a startup, you need to be a pain reliever for two different audiences: your customers and your potential acquirers. At this point, you may be thinking, "I know a startup that got acquired by a company just because . . ." and yes, we agree. Sometimes acquiring companies do buy startups for reasons other than putting an end to a nagging pain they have. But these "vitamin" acquisitions tend to have a couple things in common: they happen during frothier economic climates (like the years leading up to 2021, which broke an all-time record for US IPOs and M&As), and the resulting combination doesn't tend to do as well in the long term.

> You wouldn't try selling a product without first understanding why customers would want to buy it. Yet in our experience, startup founders regularly fail to understand what drives a company to buy another company.

There are five common pain points that companies try to solve through an acquisition: talent, product, revenue, opportunity, and profitability.

COMMON PAIN POINTS

| TALENT GAP | PRODUCT GAP | REVENUE GAP | OPPORTUNITY GAP | PROFITABILITY GAP |

Fit into one of these five, and you'll have a better chance of getting acquired — at a price that feels good — and ensuring your company continues to deliver value as part of the larger acquirer.

1. Talent gap: The acquiring company needs access to a team with a specific skill set, and your team fills that gap. This is often called an acquihire, although it can also be combined with the next pain point.

2. Product gap: The acquiring company needs a specific product feature or functionality (or technology), perhaps to match features of their competition, or perhaps to respond to a critical customer request. Your offering fills that gap. In the tech world, this is known as a "tech tuck-in."

3. Revenue gap: The acquiring company needs to grow its revenue faster, and it believes that adding your product will help it sell more of its own products.

4. Opportunity gap: The acquiring company needs to show it has a path to a larger market opportunity and has identified a complementary market segment it wants to enter. Your line of business is in that complementary market segment and can become the beachhead for the company to expand into that market.

5. Profitability gap: The acquiring company needs to improve its profitability profile. You sell your product to

the same customer base, and the acquirer believes they can maintain your revenue and improve profitability by removing overlapping costs of the combined company — euphemistically called "cost synergies."

We spend more detail on each of these five pain points and how to pitch a company feeling one of them in **Mistake 17: Selling the Future of Your Business Instead of Theirs.**

In all of these situations, you'll notice that the focus for the acquiring company is adding value to its own business, not yours.

As we said before, we often see startups approach a potential acquirer with a pitch deck and financial data about how the startup's business could grow if the acquirer bought them and used the acquirer's large sales team to sell the startup's product. That's still selling inside-out. Don't tell them what they can do for you — tell them what you can do for them.

To think outside-in, you need to develop a deep understanding of your "market landscape." Your market landscape includes your key competitors, but it also includes "co-petitors" (collaborators that could border on competitors) and potential acquirers.

> Don't tell acquirers what they can do for you
> — tell them what you can do for them.

Developing insight into what truly matters to these companies is the first step in fixing an inside-out mindset. To help you achieve this change in mindset, we've developed an outside-in checklist. The steps in this checklist will help you understand potential acquirers' biggest pain points. With this understanding, you might find yourself well positioned to solve their pain points. Or you might discover that, while you believe you're well positioned, your potential acquirers need more experience with you to believe it themselves. Or, if you really do adopt an outside-in mindset, you might become sufficiently self-aware to realize that what you're currently offering isn't what they're buying.

The Outside-In Checklist

Founders usually consider strategically developing their acquisition options in the important-but-not-urgent category — until it becomes a fire drill.

THE URGENT-IMPORTANT MATRIX

	URGENT	NON-URGENT
IMPORTANT	DO IT	SCHEDULE IT
NOT-IMPORTANT	DELEGATE IT	DELETE IT

Creating an outside-in acquisition strategy isn't something you do as a fire drill. It takes an investment in time and effort in structuring your company. Great companies will develop their acquisition strategy over a two- to three-year period, while they're simultaneously navigating their business through Pivot and Scale Waves. Doing this takes a special kind of discipline and focused thinking — the kind that can be exhausting and is easy to push to the side while remaining in firefighting mode.

There's no substitute for discipline and focused thinking. This checklist doesn't replace that, but it does provide some steps to follow and risks to watch out for.

- ✓ Identify one or two people on your executive team who are most outward looking and most long-term thinking.

- ✓ Make it their job to spend time with larger companies that might end up being acquirers.

- ✓ Include monthly market intelligence updates in your executive meetings.

- ✓ Develop and keep an updated "landscape" list — which is really a prospective acquirer list.

- ✓ Develop a strategic rationale for each target acquirer.

- ✓ Test your thesis by going back outside your four walls.

Identify One or Two People on Your Executive Team Who Are Most Outward Looking and Long-Term Thinking

In most companies, the CEO wants to be this person. Probably because it sounds glamorous — it's not. It's just hard work like every other job. If you happen to be a CEO with a great deal of experience in research, analysis, and negotiations, then you might be good for this role (and perhaps not so good for the CEO role?). Otherwise, don't fall into the trap of thinking this isn't a function, just like marketing or customer support. It's work — assign it to someone.

Your head of sales is probably most outward looking in that they are (or should be) constantly worried about whether your products and services are meeting your customer's expectations. But they're paid to be short-term focused — quarters, not years.

Your head of engineering is probably very long-term thinking. But in almost all cases, they're focused on what could be built, rather than what the rest of the world is worried about.

Small companies usually don't have the funds or the work to support a full-time head of corporate development or corporate strategy. If you have a head of partnerships that doesn't report to your sales team, they might be a good person to assign to this role. Otherwise, you'll just have to pick the next best person and support them in developing the muscle and time.

Make It Their Job to Spend Time with Larger Companies That Might End Up Being Acquirers

Incentivize these executives based on the breadth and depth of their relationships with key companies. Broad relationships mean you have many contact points at each company, so if one contact leaves to go somewhere else, you haven't lost your connection. Deep relationships mean that you develop an understanding of your target acquirer's short- and long-term priorities and challenges. Deep relationships help you start thinking outside-in, focusing on what your target acquirers need instead of what you want to sell them.

Only by creating a regular drumbeat of conversations with your target acquirers will they remember you at that specific time when something (that you won't be aware of) happens internally within their businesses that makes you an interesting potential acquisition target.

Include Monthly Market Intelligence Updates in Your Executive Meetings

It's somewhat astounding to us how rarely this is done in early-stage companies, even when someone at these companies has the relevant information. Even when they're strategically focused, executive meetings don't usually include a report on what your potential acquirers and key competitors are doing and focusing on.

Incorporating this into monthly executive meetings is one way to develop the muscle for outside-in thinking across your entire executive team. It's the repetition of focusing through the lens of your market landscape that helps your executives separate from their attachment to their internal goals and strategies. The payoff for consistently delivering these updates happens when your executive team is getting quizzed by an acquirer during the diligence process, and they're able to present the value of your startup from the acquirer's perspective instead of their own.

Develop and Keep an Updated "Landscape" List — Which in Truth Is a Prospective Acquirer List

You've given the responsibility to one or two of your executives to develop relationships with a list of companies that could possibly one day be interested in acquiring your company. They're now regularly touching base with these companies.

What happens if these very same executives leave your startup? Just as it was important to build and document processes and pull knowledge out of the heads of your SPOFs and into written materials, it's critical that your strategy executives are capturing and documenting who their contacts are, when they talk to them, what they learn, and what should be the next step in the relationship-building process. This will prevent years of work from walking out with your executive should they jump to another ship.

Develop a Strategic Rationale for Each Target Acquirer

Only now are you ready to start thinking about what you could do for them — how your company could propel each of the potential acquirers on your landscape list to the place they want to go but are challenged to, or simply don't have time to, build the machine to get there.

This doesn't mean you're looking to sell your company right now. It means you're laying the groundwork for future optionality. In this step of the checklist, you don't have to build a pitch deck — although we find going through that process is a clarifying exercise to assess whether you really know how to communicate outside-in. What's more important is that you take an honest look at your business and consider how it helps fill the gaps each potential acquirer cares to fill. Articulating this becomes the "strategic rationale" for why that acquirer would decide to acquire you. We walk through strategic rationales for each of the five pain points in the next chapter.

Test Your Thesis by Going Back Outside Your Four Walls

Use the strategic rationale you've developed to assess the likelihood that a prospect would acquire your company at some point in the future. It's better to live in the real world than to spend months or more imagining a relationship that isn't going to happen. Talk to the close contacts you've developed at these companies about your thesis. Don't pitch them unless you're looking to be acquired in the near future. But if you're still in strategy development mode, this is just a conversation: *"I've been thinking. You've been talking about your company's goal to add more capabilities to your drone delivery business in the next three years [an example of a product gap]. You know we build logistics technology. Would that help you manage more deliveries? Does that get you into any new markets you're trying to enter?"*

Many founders fear that sharing information can lead to someone else stealing their idea. That could happen, but it's more likely that not sharing will leave you in a world where you believe your own theory without testing it. Chances are, you have some incorrect assumptions that can only be vetted by sharing them. Remember, you don't know what you don't know, so be curious to vet what you believe to be true. At best, you confirm you're right. At worst, you find out you're way off base and need to go back to the drawing board. The sooner you know, the better off you'll be, and the less time you'll waste chasing a scenario that's purely fantasy.

The absolute worst-case scenario is that your contact tells you that your pitch isn't something their company would focus on either today or ever. The best case might be they like the idea enough to suggest a closer relationship.

> Many founders fear that sharing information
> can lead to someone else stealing their idea.
> That could happen, but it's more likely that
> not sharing will leave you in a world where
> you believe your own theory without testing it.

Partnering on Your Way to an Acquisition

A great deal of outside-in thinking involves developing broad and deep relationships with companies in your market landscape. One of the best ways to build these relationships is through partnerships. Partnering as a strategic path to being acquired has a number of pros, but also some cons. There's no shortage of examples of companies that partnered together first and then ended up completing an M&A.

> ⟩ Cloud accounting software company Xero acquired its ecosystem partner Hubdoc in 2018 for $70 million. Hubdoc had been a partner for four years and had invested in driving its products across the Xero customer base.[1]

> ⟩ In 2021, Intuit acquired Mailchimp for $21 billion. Mailchimp had built two integrations into Intuit's platform before the acquisition.[2]

> ⟩ Demandbase, a sales and marketing software platform, acquired DemandMatrix in 2021, just seven months after launching their partnership and sixteen months after beginning partnering discussions.[3]

> ⟩ Cisco, a serial acquirer, purchased Accedian in 2023, two years after Accedian built its Cisco-compatible product and launched it on Cisco's partner platform.[4]

---

1 Soltys, "Xero Acquires Toronto's Hubdoc."

2 "Intuit to Acquire Mailchimp," *Business Wire*.

3 Ramirez, "How Demandbase," *ELG Insider* (blog).

4 Wollenweber, "Cisco Announces," *Cisco* (blog).

Successful partnerships can be strong tailwinds as you ride your Exit Wave. The key word is "successful." Remember that many companies fall prey to the fifth mistake we discussed in the Scale Wave: treating partnerships as a side hustle. Approaching a potential acquirer to discuss a partnership, only to have that partnership falter or fail, is a fantastic way to ensure you won't get acquired by that company.

On the other hand, an overly successful partnership could stymie a potential acquisition. Acquisitions are expensive and risk-laden. If an acquirer gets everything they need from a partnership, they might just choose to continue partnering. This unexpected outcome occurs when the partnership bestows too much leverage on the potential acquirer. These stories aren't publicized because the startups are never going to admit their successful partnership crushed an acquisition opportunity.

There is one iconic example of a successful partnership and obvious acquisition going sideways. The game company Zynga relied heavily on its integration partnership with Facebook to attract new customers. Zynga grew dependent on Facebook to drive customer acquisition, which was fine with Facebook when it was smaller and looking for content for its users. But when Facebook didn't need Zynga anymore and limited its access to Facebook users, Zynga's stock price sank. It took nearly a decade for Zynga to recover its original IPO valuation, when it was finally sold to gaming company Take-Two Interactive in 2022.[5,6]

Zynga's story is an example of a partnership that's initially valuable to both companies, but the potential acquirer grows much faster, until eventually the startup's value is no longer meaningful.

There are three other common ways successful partnerships could eventually quash an acquisition:

1. The partnership drives a significant portion of the startup's revenue, so the startup can't afford to end the partnership. It has lost its leverage. Meanwhile, the potential acquirer has

---

5 Huston, "Take-Two Acquisition of Zynga."
6 "Facebook and Zynga to end close relationship," *BBC*.

one or more other companies with whom they've developed similar partnerships. If the startup walks away, the acquirer has other options.

2. The partnership is useful to both the startup and the potential acquirer. But the startup has competitors in the market that offer a good-enough alternative that may be less expensive to acquire. In this situation, the partnership just proved the strategic rationale that drove the potential acquirer to acquire its competitor.

3. The partnership proves so valuable that the potential acquirer decides to build the solution itself instead of acquiring the startup.

In all these situations, the startup's value to the acquiring company could be found elsewhere. Leaders constantly fall into the trap of believing their companies are special, unique, and sailing miles ahead of competition. When that's not the case, the partnership can simply prove this to your potential acquirers, reducing a potential valuation or eliminating a potential acquisition altogether.

Avoiding these pitfalls requires a mindful evaluation of your business strategy. In the same way that it's risky to have one customer representing a large portion of your revenue, it's dangerous to be reliant on one partner for your business success. To create optionality for your exit, you need to be relevant to multiple, ideally competing, companies, so a partner fears that losing you might enable its competition. While it might be easier to focus all your efforts on one thriving partnership, the more strategic option is to develop a few successful partnerships across competing partners.

> In the same way that it's risky to have one customer representing a large portion of your revenue, it's dangerous to be reliant on one partner for your business success.

When you do strike that partnership deal, consider what you have to offer the partnering company. Can you help them address one of the critical pain points we discussed earlier (talent, technology, revenue, opportunity, and profitability) in a way they can't eventually do themselves or get cheaper elsewhere? If you don't have an answer to that question, then perhaps you shouldn't be partnering with a view to an acquisition. At least don't until you're sure your business strategy positions you for a partnership that won't stifle a future acquisition.

Change Your Strategy to Increase Your Optionality

Unfortunately, it's not uncommon that, after you embrace outside-in thinking, you come to realize what you're currently offering isn't what many — or any — acquirers out there are buying. This is a hard thing for founders and CEOs to admit to themselves. We've seen very few achieve this level of self-actualization. Some have continued to insist, despite the evidence, that they have plenty of acquisition options, only to eventually find themselves on the nine-month countdown to running out of money. Others haven't worried about their lack of acquisition options, because their strategy has been and will continue to be to exit through an IPO. And for companies that achieve a successful IPO, there's definitely nothing to worry about. For everyone else, this isn't necessarily the end of the road — as long as you're willing, and able, to consider a change in your strategy.

A change in strategy to increase your acquisition optionality isn't something we've heard anyone talk about. It's a bold move requiring the combination of skills we describe in the Launch Wave — including becoming a futurist and not being in love with your existing solution — as well as those we describe in the Pivot Wave, since this will inevitably require some kind of pivot for your company. But this pivot might not have to be a significant one to set you on course to intersect with your potential acquirers.

The how of changing your strategy is covered in the Launch and Pivot Waves. To get a better appreciation of when it might make sense to pivot your strategy to improve your acquisition options, we offer two case studies based on real companies we worked with.

Aligning Your Product Strategy

Our first example looks at a software company that built products for software developers. In terms of our Four Waves, the company had successfully ridden the Launch Wave and, after a few pivots, was starting to see the crest of a swelling Scale Wave. They had grown to over 100 employees and were struggling to scale a better sales team, to improve their engineering velocity, and to figure out how to handle the growing number of customer service requests. Nothing unusual for this stage of a company's growth, except that leadership didn't recognize that this wasn't unusual. They took their growth challenges to be a TAM problem rather than a problem in navigating the Scale Wave. That's the easy out — blame your problems on the size of the market you chose for your product rather than on your internal execution.

To address their supposed TAM problem, they decided to develop a second product. They realized they could leverage the core of their first product to develop a product to help IT professionals. With their inside-out mindset firmly entrenched, they rationalized that entering the IT software market was a great way to expand their TAM, since that market was significantly larger than the developer software market. This strategy had two flaws.

First, the team failed to realize that selling into a different market segment required more than just a new product; it required a sales team that had contacts and could sell into IT instead of just software engineering departments. That required marketing collateral and playbooks focused on the CIO's office rather than the VP of Engineering's team. That, in turn, required different workflows and integrations with tech stacks of an IT department rather than an engineering department. The first mistake was to treat their entrance into a new market as a product initiative rather than a company initiative that required an entirely new Launch Wave attack.

> The first mistake was to treat their entrance into a new market as a product initiative rather than a company initiative that required an entirely new Launch Wave attack.

The second flaw in their strategy had to do with the company's exit opportunities. It was still years away from being large enough to consider an IPO. Their once-fast growth curve was flattening. In advising the company, we asked them to consider who their potential acquirers might be with this new product strategy and without it. Without the strategy, there were a number of companies that might have been high-value acquirers, including large developer tool companies. And there was no shortage of IT software tool companies. But there were very few companies that sold both developer tools and IT tools, and they tended to be behemoth companies like Microsoft that didn't need the solutions this startup was offering.

By expanding their product offering across multiple different segments, the company was limiting its acquisition optionality.

The company did eventually get acquired — by a software tools company. The acquirer attributed no value to the IT solution part of the business because it solved no pain point for them.

Perhaps this is a case study of increasing your acquisition options by not changing your strategy. But looking deeper, it's also a lesson in incorporating acquisition optionality into your overall product strategy to help you better decide how to set your business course.

Aligning Your Business Strategy

Our second case study looks at a software startup that had successfully grown to over $100 million in annual recurring revenue through a product that had an immediate product-market fit and a CEO who managed to build a world-class sales team. This was a company that had the potential to IPO. So why worry about building out some acquisition options? What could go wrong?

As it was successfully riding its Scale Wave, the startup was also doing the right thing and exploring partnering opportunities with other, much larger, companies. One company in particular seemed to be interested in a significant partnering opportunity. This was definitely a potential acquirer. The startup's product filled a technology gap in the acquiring company's product portfolio, a gap that was significant because it was where the market was heading, rather than where it had been. The two companies talked for many, many months (because strategic transactions take time). Eventually, the larger company put forth a proposal. They wanted to white label the startup's product into their technology portfolio and sell it through their own sales team.

> The upside: The larger company had a sales team at least four times the size of the startup's sales team, and at least ten times more customers than the startup had. The opportunity to scale through this partnership would be massive. And after some negotiation, they agreed to keep the startup's product name on the product they sold, so the startup's brand awareness would grow.

> The downside: The startup's sales team would be limited in what they could sell directly going forward. The market would be split with the potential acquirer selling into one segment, leaving the startup's sales team with the traditional segment they were already selling into. But the company prided itself on its world-class sales team. And the team itself obviously hated this idea because it limited their influence in the business.

The startup walked away from the opportunity. Just having the opportunity made it feel as though it was going to have no shortage of optionality. And it was still gunning for an IPO. Unfortunately, the company didn't perform an outside-in analysis. Had they done so, they would've come to realize this partner was the only large company in the market landscape that was looking to buy what they had to sell. The existence of this opportunity wasn't evidence of anything more than one company's interest.

What happened next was 2022 — a year that was a Rogue Wave if ever there was one for pre-IPO companies. The company had to delay its IPO because the markets tanked. An analysis performed by Pitchbook showed that there were fewer exits in 2022 than there were in any of the nine previous years and the value of those exits is the lowest it's been in a decade.[7]

The down-market hit the potential partner also. Had the startup been willing to align its business model earlier, the potential partner might have viewed the startup as a good acquisition opportunity to bolster its business through the rough market.

At the time of writing *Sail to Scale*, the startup's story continues. Like most software companies, they took a hit in growth through 2022 and 2023, but they continue to plug on. The IPO markets are still closed to them. And now they know they have few acquisition options with their current strategy. In the end, this company will find its way to an exit of some sort. But the business strategy decision they made added years to their journey.

When you embrace outside-in thinking, you may come to realize that what you're currently offering isn't what many — or any — acquirers out there are buying. If that realization hits you, you have a hard choice to make: either go it alone or pivot your strategy to better fit within the plans of potential buyers. These two case studies are cautionary tales of what happens when you ignore the signals potential buyers are sending you.

---

7 Pitchbook, *PitchBook-NVCA Venture*, 37.

Key Takeaways

1. There are five common pain points that companies try to solve through an acquisition: talent, product, revenue, opportunity, and profitability. Fit into one of these five, and you'll have a better chance of getting acquired at a price that feels good.

2. Great companies develop their acquisition strategy over a two- to three-year period while they're simultaneously navigating their business through Pivot and Scale Waves.

3. A successful partnership is an opportunity to build strong connections with a potential acquirer. But there are times a partnership is too successful, and the prospective acquirer doesn't see any additional value in owning you.

4. Sometimes, it takes a bold change in strategy to increase your acquisition optionality. This can require a change in your product strategy to fit within the product lines of more potential acquirers. Or it might require a change in business model strategy to be able to partner more closely with a potential acquirer and demonstrate your value to them.

Mistake 17: Selling the Future of Your Business Instead of Theirs

Even when a startup founder learns how to think outside-in, they often seem to forget everything they've learned when they finally get a meeting with a potential acquirer. Instead of pitching how their business can solve an acquirer's pain point, or even how their business can be a "vitamin" to help the acquirer's strategy accelerate, startup founders and CEOs in love with their own business pitch how their own business can grow with the acquirer's help.

It sounds crazy. And yet we'll be sitting in a pitch meeting and thinking, "Yeah. That happened. Again."

The most common theme in these failed pitch sessions is the "Be My Sales Team" refrain. The startup has found its product-market fit but is having challenges scaling. They believe their scaling challenges are due to not having enough money to hire more salespeople. If only they had more money, they'd have more salespeople who would sell more of their product, so they could earn more money. The CEO finally gets a meeting with a potential acquirer that, perhaps, sells into the same user profile, sells a complementary product offering, or sells in an adjacent market. At the pitch meeting, the CEO explains how great it's going to be when the acquirer's sales team adds the startup's product to their sales bag. Because this will drive more of the startup's product revenue. More revenue for the startup, more revenue for the company that acquires it. Right?

The problem with this pitch arises from the startup selling the future of its own business rather than the future of the acquirer.

The acquiring company has a business it has invested years of time and money building. It's trying to tame its own Scale Wave, and to do so well, it needs to focus on investing in what sells more of its own products, rather than diverting its sales team to sell more of a different product.

Sell What Your Acquirer is Buying

Mona recalls her time as a Corporate Vice President at Cadence Design Systems running their global M&A activities. Her children were young at the time and didn't understand what she did. She told them that she bought things for her company. Mona's youngest son asked if she bought things like shoes. "No," she replied, "I buy other companies, not specific products." He thought for a minute and then said, "Your job is to buy companies. And you like buying shoes. So why don't you buy a shoe company for Cadence?"

It's a cute story. But more than that, Mona remembers his comment every time she hears a pitch from a startup suggesting she should buy their company to sell their product. She's paid to figure out ways to sell more of her company's own products. "If selling their product helps us sell more of our product, then tell me how," she suggests. "Otherwise, your pitch sounds like my young son's pitch suggesting a software company buy a shoe company."

When you finally do get that meeting with an acquirer, you want to position yourself to be selling what they're buying. That means framing (or reframing) your pitch to show how your business fills one of the acquirer's gaps.

Understanding Your Acquirer's Decision-Making Process

Companies with experienced acquisition executives on board will build their own internal business case for each acquisition. Large companies that are active acquirers (called serial acquirers) often have acquisition review committees that review the business case prepared by the internal sponsor of the acquisition and the corporate development team.

The business case, or acquisition review deck, will include the following sections:

> Summary of the target company

> The internal business sponsor

> The thesis for the acquisition (what we call the "strategic rationale")

> The assumptions made in developing the strategic rationale

> Alternatives to this acquisition

> Risks to achieving the strategic rationale and possible mitigations

> Effect on the acquiring company's financials

The best acquirers also include a summary of the key integration actions that need to occur to achieve the desired results from the strategic rationale.

If you're talking to a potential acquirer who hasn't done many, or any, acquisitions in the past, then they might not have a well-defined decision-making process set up. If that's the case, it doesn't bode well for a successful acquisition. Putting that aside, you'll still need an internal sponsor, whether the acquirer calls it that or not.

As a company in potential exit discussions, you need to know who your internal sponsor is, and how much of a champion they are for your business. You also need to understand their decision-making process. Just ask your contact.

If it doesn't sound as though you have a strong sponsor, then go back to **Mistake 16: You're Inside-Out** and do the hard work of getting to know your acquirers better. If you have a sponsor, then help them by making their internal business case as easy as possible to build.

Making Your Sponsor's Life Easier

The individual who acts as your sponsor within the acquiring company is your lifeline to a successful acquisition — really to any successful corporate transaction. It sounds trite to say that your job is to make sure your sponsor can easily tell your story on your behalf. And yet, we've seen CEOs get frustrated with sponsors who push back on the value of their business. Even if you aren't interested in selling your company — even if the acquiring company is coming to you with an acquisition proposal instead of you coming to them — there's no situation in which it makes sense for you to express frustration at them. Because you never know when you'll need their help or what role they'll end up having the next time you need it. Don't turn away an offer before you've received one.

Don't turn away an offer before you've received one.

If your sponsor believes that acquiring your startup makes sense, they must then convince their company. Don't leave your sponsor to do this on their own. Help them sell the acquisition. Congratulations! You've been promoted from business development representative (thinking outside-in) to sales representative.

As the sales rep for your business's potential sale, you understand your sponsor has a full-time job of their own. And since you also have your own full-time job of helping run your company, you should realize that any number of competing priorities might pull your sponsor's attention away from your transaction, leaving you in what Dr. Seuss aptly called "the waiting place."[8] The easier you make your sponsor's job, the easier yours will become as well.

Checklist for Making Your Sponsor's Life Easier

1. Make sure you understand the acquiring company's decision-making process. If they don't have one, be bold and suggest a process for your sponsor to propose to the company. Work with your sponsor to fit the process into the company's decision-making culture. If your sponsor delivers a solid decision-making process to their CEO, it makes your sponsor look good — which normally makes them more aligned with you.

2. Whether or not the acquirer's decision-making process requires a written business case, write one for your sponsor. The format of the business case should align with the format preferred by the acquiring company. Some companies work entirely in slide decks. Others prefer the Amazon six-pager approach. The substance of your business case should include the "strategic rationale" for the acquisition. We'll explain more about strategic rationales in the next section.

3. Keep in touch with your sponsor regularly to stay up to date on the internal process. A good salesperson doesn't wait for the customer to get back to them. They call — enough times to make the customer feel special but not so many times that the customer is creeped out. You need to find the same balance. Don't just reach out to say hi. Find out where the process is. Ask how the sponsor is feeling about it. Offer help when you can.

---

8 Seuss, *Oh, The Places You'll Go*.

4. **Don't forget to give as well as get.** No business relationship should be one way. The acquisition process takes time. If during that time you can continue to partner with the acquiring company and help it advance its strategic goals, it goes a long way to adding credibility to the overall business case.

Heather recalled that during the acquisition process when she was the CEO at Hitch, she learned in a conversation with her sponsors at ServiceNow that they needed help selling the acquisition to the head of engineering. Heather wouldn't have known this was a blocker had she not kept in touch and developed a close business relationship with her sponsors. Finding this out early gave her an opportunity to give them information specifically relevant to their engineering organization to help them better articulate the benefits within that team.

How to Create an Effective Strategic Rationale

The strategic rationale is a key part of the acquirer's business case for why they want to acquire your company. As we said above, if your potential acquirer isn't a sophisticated acquirer, then they might not be preparing a formal business case or strategic rationale. That's not so great for them. But you can use that to your advantage and formulate a strategic rationale for your sponsor to pitch internally that puts you in the best light. And if your potential acquirer does prepare a more formal business case, you now know to reach out to your sponsor and offer to help them prepare their strategic rationale.

An acquisition strategic rationale is the thesis for why the acquirer wants to acquire the target company. Almost all strategic rationales fall within one of the five common pain points companies try to solve for through an acquisition. These five pain points are detailed next, along with what could make you a good fit within each one.

Pain Point 1 — The Talent Gap

Sometimes a company will decide to acquire a startup in order to bring the startup's team into their own ranks. This is an example of an acquihire, and it normally also includes some technology or content the team developed during their time at the startup. For this kind of acquisition to be a meaningful exit, the team has to be something special. Here are three examples of acquihires we've worked on with different outcomes.

1. A three-person company, led by Ph.D. experts in advanced mapping using millimeter wave radar, developed some technology that was obviously relevant to a number of large companies. The technology itself was valuable, but it hadn't been commercialized yet. The team had been working on it during their Ph.D. program and for another year in their startup. The opportunity for the team to build out a product in their area of expertise as part of a large enterprise was enticing. And the acquiring company didn't have expertise in this area and needed it to achieve part of its strategy. The three-person startup had only raised some friends and family funding, so the vast majority of the substantial acquisition price went to them.

2. Another three-person startup had raised a small amount of seed funding and used it to develop a product that half a dozen customers were using. However, they were running out of money and weren't able to get more customers to sign up and pay for the product in its current state. This startup thought they'd found their product-market fit, when in reality, they'd found a few customers in the innovator category who were willing to pay a little bit to try out what they'd built. They managed to negotiate an acquihire because the employees' expertise aligned with the acquiring company's needs, and the acquirer was growing quickly and had several open roles to fill. But unlike the prior example,

the acquisition price was only sufficient to cover the company's debt obligations.

3. Not all acquihires are three-person teams. And not all of them are publicized as acquihires. In January 2024, a startup called Einblick that combined data analysis with generative AI announced it was acquired by cloud data company Databricks, a $43 billion market cap company.[9] While not announced as such, this would be categorized as an acquihire along with the purchase of the technology the dozen or so employees had developed. How did a tiny startup get the attention of a multi-billion dollar company? For one, the founders of both companies went to college together.

Acquihires are usually only an option for a small team with a complementary expertise that can come in as a unit into an acquiring company. There are always exceptions. Consider the "almost acquihire" of more than 700 employees of OpenAI in late 2023 after the CEO of OpenAI, Sam Altman, was ineptly "fired" by its board of directors. It was a dramatic weekend for those in the world of tech, as shocking news about the company that created ChatGPT kept pouring in seemingly by the hour. First was Altman's firing on Friday, November 17, 2023, and then a threat by the employees of OpenAI to quit if Altman wasn't reinstated. Next came an open offer from Microsoft to hire any OpenAI employee that wanted to resign. Five days later, Altman was back in his CEO seat, the board was practically disbanded (later to be replaced), and the almost acquihire of the century by Microsoft didn't come to be.[10]

What to take away from all of this? In the early stages of your company's life, the value is in the team and its differentiated expertise. It normally takes a few years to add product value to your startup, and the path is fraught with risk. If you don't make it through your launch phase and an acquisition is your best path, it's your team that drives any value you might have.

9 Azhar, "Databricks Acquires Einblick Team," Databricks.

10 O'Brien and Hadero, "OpenAI brings back Sam Altman."

> In the early stages of your company's life, the value is in the team and its differentiated expertise.

Many startups outsource the work of creating their MVP to freelancers instead of members of a founding team. Whether that MVP is a shoe design, a software app, or a new health drink, the core of the product should be built by the founding team. No one starts a new business with the plan of exiting through an acquihire. But if that's where you end up, you'll only get value because the acquiring company sees your team as a better path to getting the talent they need rather than by hiring people individually. The value you command increases if you can show the acquirer that your team has demonstrated the ability to work together to solve complex problems and create elegant solutions, that they have the expertise in an area critical for the acquirer, and that they're willing to continue working together as a team in full-time roles at the acquiring company.

Pain Point 2 — The Product Gap

More often than using an acquihire to fill a talent gap, companies acquire startups to fill a gap in their product offering. To understand the strategic rationale of your acquirer, it's important to understand why the acquiring company believes they have a gap. Is it because their competitor has a feature they don't have but their customers are demanding? Because that sounds like a pain point that needs to be filled. However, if the gap is something the company would like to fill to expand their offering, but it isn't a threat to their current revenue, then you're in danger of being in the "nice-to-have" acquisition category, putting you in a much weaker leverage position. Acquisitions to fill product gaps are more often justifiable when they help the acquirer stop losses of existing revenue, rather than drive gains of new revenue.

> Acquisitions to fill product gaps are more often justifiable when they help the acquirer stop losses of existing revenue, rather than drive gains of new revenue.

If you've done your acquisition homework, then you've developed your business strategy toward that product feature or functionality that one or two large companies in your market landscape haven't built. Yet what if their equally large competitor has not only built it, but also has positioned that feature or functionality as a key differentiator to their common customers? There's no better position to be in as a startup. If you were strategic enough to navigate yourself into that position, then make sure you make it clear in the material you send to your sponsor to support their strategic rationale.

Even when you appear to be the perfect medicine to solve their pain point, experienced acquirers will know that acquisitions most often fail when the acquiring company doesn't successfully integrate the target's product into their overall product line. This is especially true for technology acquisitions, so your pitch to the acquirer and your write-up for your sponsor must emphasize the following.

> What you've learned from your market research about the customer demand for a product line that includes your feature or functionality: Include both quantitative data and qualitative feedback from actual customers, either of the acquiring company or its competitor (not necessarily your customers).

> How your team can help the acquiring company compete better: Ideally, the answer will be because you have a depth of experience in supporting and selling your feature/functionality into their overall solution already.

> What you've done to make it easy for the acquiring company to integrate your feature/functionality into their overall product line: If you're a software company that has done its acquisition homework, then you'll point out that your software is written in the same language as your acquirer's, and that you've built APIs to easily integrate your software into anyone's architecture (your acquirer's or their competitor's).

> The time it will take to integrate your feature/functionality into the acquirer's product line and bring it to market: This point combats the prevalent not-invented-here (NIH) mindset of the acquirer's engineering team. The goal is to show that integrating your feature/functionality is much faster than having the acquirer's engineering team try to build a competing version in-house. Your aim is to convince the acquirer that buying you is de-risking revenue loss, compared with taking a chance on expending their own internal development resources to build an equally or better-functioning solution within a reasonable time.

On the other hand, if you determine from talking to your sponsor that you're in the "nice-to-have" category, be wary of the amount of time your sponsor takes up from you without seeing broader company commitment. Time kills all deals.

You could be a much smaller company than the acquirer you're negotiating with, but your time is just as precious as theirs, and probably more so since large companies tend to have more redundancy in their ranks. A person with a senior title at a large company can spend a lot of time meeting with you and can even engage others to send you NDAs and do some diligence on your business. Their investment in working with you is flattering. The time they spend doesn't affect their business much, but it's potentially preventing you and your team from moving your business forward.

> You could be a much smaller company than the acquirer you're negotiating with, but your time is just as precious as theirs, and probably more so since large companies tend to have more redundancy in their ranks.

A machine learning platform startup we know had this very experience. A Fortune 50 tech giant took up more than six months of their time, even signed an acquisition term sheet with them, before pulling out of the deal late in the negotiations — simply because there wasn't broad-based interest in the acquisition. Ultimately, the startup was seen as a "nice-to-have" by enough executives that the sponsor couldn't push it over the line, although they certainly did try. The startup had a hard time recovering because they'd spent so much time with the potential acquirer, their entire small team had become aware of the acquisition opportunity and been distracted for more than half a year. Luckily, the two co-founders had built a loyal and engaged culture, and the startup was able to rebound from the experience. Others haven't been as fortunate.

Pain Point 3 — The Revenue Gap

If your company is larger and has managed to generate revenue that's meaningful for the acquirer you're talking to, then they may be interested in your startup in order to fill a revenue gap.

To build a strong pitch for a revenue gap strategic rationale, demonstrate the following:

> The acquirer will be able to maintain and grow your revenue. This is best demonstrated by showing that you have strong gross and net retention.

> Your business will drive more of the acquirer's own product revenue.

> You can do the above without the acquirer having to provide much additional investment in your business.

> Your business can be easily integrated into the acquirer's. Showing that your product's average selling price is in the same range as your acquirer's products, that your pricing model is similar to your acquirer's, and that you sell to the same kind of user as your acquirer are all great indications that integration will have the opportunity to succeed.

Gross retention is the percentage of recurring revenue retained from existing customers, inclusive of the effects of downgrades and cancellations. It reflects your ability to retain your existing customers and is below 100 percent.

Net retention is the percentage of recurring revenue from existing customers, inclusive of upsells, cross-sells, and add-ons as well as the effects of downgrades and cancellations. It reflects your ability to grow revenue from your existing customers and is typically above 100 percent.

If your acquirer is buying to fill a revenue gap, then the pre-diligence and diligence will be focused on your customers, your financial metrics, and your financial forecast. Technical founders and CEOs often make the mistake of focusing on the details of their product and technology in discussions with acquirers. Product details matter less to a revenue-focused acquirer. Too much focus on the product, to the exclusion of detailed customer and revenue data, can make it look like the target company is either trying to hide something or just not business savvy.

During diligence, an acquirer will want to see a bottom-up build of your revenue forecast along with the assumptions you're making. If you have a strong CFO (and a solid business opportunity), this shouldn't be hard to build. Otherwise, you may want to consider hiring an M&A advisor to help you prepare.

Pain Point 4 — The Opportunity Gap

The story of CMEC at the start of this section is one about an acquisition that filled an opportunity gap for the acquirer. The many companies that were interested in acquiring CMEC were interested in expanding their operations into more states to expand their market, and CMEC was operating in the very states in which they had gaps.

Acquisitions intended to expand geographically are common, but aren't the only opportunity gap transactions. Companies might seek to expand their market opportunity by acquiring companies that sell to other types of customers, such as Coca-Cola's acquisition of energy drink business Monster for over $2 billion.[11] This was meant to expand Coca-Cola's reach to people who may not drink soda but do like their caffeine hits.

Other companies might seek to expand into a new market by combining their business with a target company's. Opportunity gap acquisitions are probably the hardest to pull off successfully. This category is particularly difficult because you're trying to take two companies that have each done one thing (hopefully well, for many years) and make them work together while also learning how to do something new, and while not losing each of their current businesses along the way. Broadcom's acquisition of VMware in late 2023 might be an example of such an acquisition. Broadcom, a semiconductor company, acquired VMware, a software cloud computing company, with the intent of combining the two business opportunities to offer hybrid public/private cloud hardware and software services to customers, thereby entering a new market dominated by the likes of Microsoft and IBM.[12]

Acquisitions to fill opportunity gaps tend to be directed toward larger companies rather than startups. It's rare to see early-stage companies creating meaningful new market opportunities for large acquirers. And if a smaller company is looking to acquire to create a new opportunity, that generally signals a problem with

---

11 McGrath, "Coca-Cola Buys."
12 Vogel, "Broadcom's acquisition," SoftwareOne.

the acquirer's own business, requiring it to find a different market in order to grow.

> Acquisitions to fill opportunity gaps tend to be directed toward larger companies rather than startups. It's rare to see early-stage companies creating meaningful new market opportunities for large acquirers.

With that warning, if you're in discussions and determine your acquirer is looking for you to fill their opportunity gap, then your focus should be on showing your strength within that opportunity. Like CMEC, you'll want to focus on the strength of your customer relationships and your sales force within the relevant market. The acquirer won't have much experience selling into this new market. They'll have to rely on your sales force to both continue selling and efficiently train their sales team, so they can help grow sales in this market.

Pain Point 5 — The Profitability Gap

The final gap that drives a company to acquire another company is the profitability gap. Like the opportunity gap, the profitability gap is most often a strategic rationale applied to larger acquisition targets. The acquiring company believes the target company is both complementary to its business, and, just as critically, that it will help it improve its profitability profile.

If you've managed to ride the Scale Wave successfully, and your financial metrics are sound enough that you don't need additional outside investment to continue your business, then you start looking like a prime candidate for an acquisition. As long as you've made sure your business strategy has aligned with the strategic plans of at least two or three larger companies in the market landscape, you have probably set yourself up for some excellent exit optionality.

Your pitch to the acquirer and your write-up for your sponsor will focus on three main points:

1. Why your business is complementary to the acquirer's business

2. Whether there will be "revenue synergies"

3. Whether there will be "cost synergies"

1. Why your business is complementary to the acquirer's business

We're no longer talking about your technology or product here. Acquirers buying to fill this gap want to know about your customer retention and pricing model. They want to know whether you're selling to the same customer profile they are at the same or a similar price point, so your business model is complementary to theirs.

2. Whether there will be "revenue synergies"

Revenue synergies are created when the result of the acquisition drives more revenue than the sum of the revenues of the target company and the acquiring company. Revenue synergies are tough to create. Nearly every M&A financial projection we've seen has shown revenue synergies, and yet very few have achieved them. A survey conducted by McKinsey & Company reported that companies consistently fall short of their aspiration for revenue synergies, with an average gap of 23 percent between goal and attainment.[13]

As a startup looking to get acquired, why not create a compelling pitch for revenue synergies, and perhaps deal with the reality of achieving them later? This could backfire on you if, during your term sheet negotiations, the acquirer demands that part of the purchase price for your company is paid as an earnout.

---

13 Chartier et al., "Seven rules," McKinsey & Company.

> Earnout is a term in the acquisition term sheet allocating some of the purchase price for the acquisition to be paid in the future, after the acquisition has closed, and only upon the acquired business achieving certain financial goals.

We'll discuss earnouts more at the end of this chapter. For now, let's focus on how revenue synergies are most often created by using Microsoft's acquisition of Activision as an example.

Activision has been one of the most successful video game companies in history, growing to over $7 billion in revenue due to top game series like *Call of Duty*. Buying Activision enabled Microsoft to expand its reach to Activision game die-hards, cross-selling other Microsoft products (such as Microsoft's Game Pass subscription), and driving more Xbox sales.[14] The increased sales of Microsoft products to Activision customers were the revenue synergies created by the acquisition.

Showing that your acquirer can cross-sell their products and services to your customer base is the strongest way to show revenue synergies. Other revenue synergies may result from selling your product to your acquirer's customer base. Although, as we've mentioned throughout this Rogue Wave, your acquirer is usually more concerned with growing their business than growing yours.

> Showing that your acquirer can cross-sell their products and services to your customer base is the strongest way to show revenue synergies.

Finally, sometimes revenue synergies are created because the acquisition creates a leadership position that allows the combined company to increase prices because there are few other options left for customers. Sometimes, though, this strategic rationale will

---

14 Microsoft News Center, "Microsoft to acquire," Microsoft.

be challenged by anti-trust regulators as being anti-competitive. So best not to be pitching that as the basis for your strategic rationale.

3. Whether there will be "cost synergies"

The corollary to revenue synergies are cost synergies, which are created when the result of an acquisition can reduce the cost of operating both companies together compared with operating them separately. This occurs by taking advantage of overlapping operations (such as redundant offices), consolidating enterprise software systems, and reducing the workforce due to duplicate roles. If the combined company can keep the combined revenue and reduce the combined cost, then profitability increases.

Cost synergies are often a euphemism for conducting a layoff after an acquisition. This is one of the hardest periods for a leader to manage, as their years of pursuit to build a sustainable, scalable business with a great team and culture now turns into a financial engineering exercise. Some of the people that have been with you along the journey are the very people you have to fire as part of the acquirer's path to increased profitability.

A Quick Word on Earnouts

If you're far enough along in your acquisition discussion to be talking to a potential acquirer about the value they'll gain by acquiring your company, well first — good for you! You've made it farther than most. And second, don't get too self-confident.

It's easy to build financial forecasts that show how much more revenue the acquirer will be able to capture, either through their business or yours, as a result of the value you deliver (the synergies we discussed above). But if you're not in a bidding war between two or more buyers, you might find your acquirer asking you to put your money behind your words. They do this by suggesting that part of the purchase price be paid out as an earnout.

For an acquirer, an earnout helps eliminate uncertainty or risk associated with future performance after the deal closes.

Instead of paying, say, $50 million up front, the acquirer might suggest paying $30 million up front, and another $20 million over a two-year period if the business achieves certain milestones that prove out the synergies you claim. The near-perpetual problem with earnouts is twofold.

First, while you're building your revenue forecasts for the combined business, you'll be suffering a severe case of optimism bias — the tendency to overestimate the likelihood of experiencing positive events and underestimate the likelihood of experiencing negative events. In this delirium, you'll come up with persuasive reasons why your $10 million in annual revenue, when combined with the acquirer's $100 million in annual revenue, will, within two short years, build up into an exhilarating $400 million in annual revenue. The acquirer might then suggest that 40 percent of your purchase price be paid if the combined company reaches that $400 million annual revenue target within two years.

This ties in nicely with the second perpetual problem — even the best laid plans go awry. And that's especially true if it's a business plan to be realized over more than a year. Even if you weren't overly optimistic in the forecast that resulted in the earnout structure, the acquirer might end up changing how it carries on the business in a way that makes it difficult for you to achieve your earnout targets. This situation happens more often than not, leading target company employees to feel "robbed" of their earnout.

Don't forget to consider negative revenue synergies. Negative revenue synergies result when the combined revenue of the target and its acquirer after an acquisition is less than before the acquisition. For example, negative revenue synergies can occur if one company was a customer of both the acquirer and the target company before the acquisition. After the companies merge, the customer demands that the two companies' products be bundled at a discounted price (or even for free).

Agreeing to an earnout is like tying your dinghy to the back of your acquirer's luxury ship, jumping on board, getting wined and dined, and agreeing to set sail for a small island in the middle of the ocean. But if the ship doesn't make it there on time, you have to go back and sleep in your dinghy for the rest of the trip. And if a rogue wave hits in the meantime, you could be the one who drowns.

Key Takeaways

1. As a company in potential exit discussions, you need to know who your internal sponsor is and how much of a champion they really are for your business. And you also need to understand their decision-making process.

2. An acquisition strategic rationale is the thesis for why the acquirer wants to acquire the target company. Almost all strategic rationales fall within one of the five common pain points companies try to solve for through an acquisition: a gap in talent, product, revenue, opportunity, or profitability.

3. In the early stages of your company's life, the value is in the team and its differentiated expertise. If your acquirer is filling a talent gap, they see your team as a better path to getting the talent they need than by hiring people individually.

4. If your acquirer is filling a product gap, then make sure your pitch explains how you've made it simple to successfully integrate your product into their overall product line.

5. If your acquirer is buying to fill its revenue gap, then the pre-diligence and diligence will be focused on your customers, your financial metrics, and your financial forecast.

6. If your acquirer is filling an opportunity gap, then your acquirer won't have much experience selling into this new market. Focus on showing the strength of your customer relationships and your sales force within the relevant market.

7. If your acquirer is filling a profitability gap, then they'll want to see both revenue synergies and cost synergies.

Sidebar:
The Pre-Diligence List

Before an acquirer is ready to present you with a term sheet, they'll often want to do some pre-diligence. Where they spend their time should be a signal of what they consider most valuable about your business. Full-blown diligence, which occurs after a term sheet is signed and before the acquisition can close, is a hair-raising experience for all business leaders and usually involves hundreds of data requests. Its precursor, pre-diligence, can include discussions and review of data on the following topics.

General Business Topics

> Business overview presentation
> Business models (licenses, royalties, subscriptions, product sales, maintenance, support)
> Capitalization table
> Leadership team and board
> Key contracts other than commercial/sales contracts
> Key competitors, what they do better than you, and what you do better than them
> Where you're seeing the most demand for your business and why

Technical

> Tech stack/infrastructure
> List of products/service offerings
> Supported standards, certificates, and qualifications
> List of tools used in development flow
> Details of any outsourced product development work
> Any export restrictions
> Patents

Customers

> Top ten customers (without providing names) and what percentage of annual revenues they make up in the aggregate

> Packaging options, list price, and average selling price

> Overview of any distributor/seller relationships, by product and/or location

> Customer support services and infrastructure

> Percent of revenue from new customers versus expansions in existing customers versus flat renewals

> Large past customers that are no longer customers

Finance

> Profit and loss statements by quarter (two years back and two years forecast)

> Revenues by product line by quarter (two years back and two years forecast)

> Average collection period of accounts receivable

> Operating expense by function

> Capital expenditures, if your business is a capital-heavy business

> Prior fundraising and financings

Employees

> Organizational structure

> Headcount (without employee names), by title, function, product line, length of service, location, and salary and benefit costs

> Total headcount by quarter (two years back and two years forecast)

> Benefits plan (high-level description)

> Compensation plans (high-level description)

Mistake 18: Rejecting the Right Acquisition Offer

We're almost at the end of this journey. But not the end of your journey — you can go on to ride your Scale Wave or navigate your Exit Wave to join a bigger ship. And then perhaps you can launch more businesses within that larger environment as an intrapreneur. Or scratch that entrepreneurial itch and leave to launch a whole new business.

We hope the mistakes we've seen and made have helped you get to this point in your journey with many options available to you. If we've managed to help you navigate to an acquisition offer, then you now have an interesting option to consider and an important decision to make. And one more mistake to avoid.

Some acquisition offers should not be accepted. But more offers should not have been rejected. Companies generally don't publicize their rejection of acquisition offers. But we did manage to find a few stories to share.

> Some acquisition offers should not be accepted. But more offers should not have been rejected.

The Ones That Got Away

Groupon is an e-commerce marketplace that allows you to purchase vouchers for products at discounted prices. It launched in 2008 and went public in 2011, at a market cap of $17.8 billion. That sounds like the perfect exit if there ever was one, and certainly it made a lot of people a lot of money. But was staying separate as an independent company the right decision for Groupon?

Just two years after it launched, Google offered to acquire the company for $5.75 billion. Groupon rejected the offer — and that decision looked pretty good for a while. But even at the IPO, there were concerns. Jim Cramer from *CNBC* called the Groupon IPO "the most hyped, most artificial deal I've ever seen since the dotcom era began."[15] Less than two weeks after its debut, shares tumbled, and it's been a bumpy ride ever since. The founder-CEO was fired in 2013. The co-founder took over as CEO and then was removed in 2015. And they've had more changes since. At the start of 2024, after operating for fifteen years, its market cap was hovering down around $500 million, making that $5.75 billion offer from Google over a decade ago a distant dream.[16, 17]

In another example, D2iQ was launched in 2013 under the name Mesosphere, to make it easier for developers to manage their applications in a data center. In less than two years, they raised $49 million in venture capital from some of the biggest names in the industry, including Andreessen Horowitz and Khosla Ventures. In 2015, they received a $150 million offer from Microsoft, which they believed (or perhaps their investors believed) was too low. And so they turned it down. They continued to try scaling their operations, raising another nearly $200 million in the process, and pivoting their business, trying to find a path to growth that matched the expectations of the money they'd raised. But by 2020, they were quietly searching for an exit. The problem: the amount of venture capital they'd raised made them very expensive to acquire. In the end they didn't find a buyer, and at the end of 2023, they announced they were shutting down.[18]

---

15 Lyons, "Groupon IPO No Bargain as Early Investors Milk the Company."
16 Gustin, "Groupon Fires CEO Andrew Mason."
17 Mac, "Groupon Shares Crumble After Company Names New CEO."
18 Weinberg, "Software Startup That Rejected Buyout From Microsoft" *The Information.*

Investors and Acquirers Like the Same Businesses

There's an understandable reason startups reject acquisition opportunities. Companies rarely come knocking on the door of a struggling startup, excited about acquiring them. Like investors, acquirers like to invest in businesses that look like they're doing well.

But when your business is doing well, you feel really good about things. And optimism bias kicks in. Remember our Startup Valuation Curve in **Mistake 15: Running Out of Runway**? It bears repeating. It turns out that your odds of raising money are highly correlated to your odds of getting acquired. Companies like Groupon and D2iQ were getting acquisition offers at the same time they were getting interest from investors. With all this attention aimed at them, founders become overconfident, and acquisition offers rarely hit the premium founders feel they deserve given their current success. And so they walk away.

> Your odds of raising money are highly correlated to your odds of getting acquired.

Earlier, we mentioned that Hitch ended up getting acquired by ServiceNow in 2022. At the same time the acquisition offer came in, Heather was finalizing a funding round. Heather remembers struggling to manage two different diligence reviews at the same time. It was the hardest thing she has ever done in her career. The stakes were so high; it felt like competing in the Super Bowl.

Why did she, and her board, choose the ServiceNow offer over investors? She saw the skills intelligence technology was core to what ServiceNow wanted for their business. That meant her team would have a home in a mega-cap company invested in making the Hitch product part of their overall portfolio. Today, Heather is leading the data science team in delivering next-generation AI solutions and workforce intelligence products that feature Hitch's technology, which has been rebuilt into the ServiceNow platform for a seamless experience.

Think Like an Entrepreneur, Not an Investor

Like all startup curves, the lifecycle of a startup goes from conviction and enthusiasm to reality and disillusionment and then back again many times during its growth. At a company's high point, no one wants to sell. But at a company's low point, no one wants to buy. So when you're riding high, how do you avoid the mistake of rejecting a good acquisition offer?

The answer depends on what you define as success (which is not the same as what your investors define as success). This brings us full circle to the start of this section, which explained how venture capitalists don't like it when their startups talk about preparing to be acquired. That's because the business model of most venture capital firms requires them to build a portfolio of investments in companies that swing for the unicorn — or these days, the decacorn — fence and grow into public-company-worthy businesses. The venture capital business model needs this because odds are only a select few of their portfolio investments will make that leap, meaning all the return they get must come from a small handful of investments. Which handful will succeed is pretty much impossible to tell at the beginning, so all of them have to try to get there, or else VCs might miss out on the one startup that could've made it.

That's the venture capital business model. And it's a rational one for them. But we have a saying: you're an entrepreneur, not an investor. So think like an entrepreneur, not like an investor.

For some entrepreneurs, the definition of success is completely aligned with that of their investors. They'd rather end up with nothing than sell early and miss out on the less than 3 percent possibility that they might build a billion-dollar valued company. If that's you, then it's not a mistake to reject an acquisition offer that doesn't meet your billion-dollar standard.

However, most of the entrepreneurs we've worked with, worked for, and advised do not define their success in that way. Many care deeply about the product they're building and want that product to be used and to deliver value broadly in an industry. For those founders, finding an acquirer with a complementary offering, a large distribution channel, and a strategic plan that requires their product is the most likely way to achieve success.

Other founders and leaders place a priority on the employees that have helped them grow their business. They care deeply about the team they've built and want to do everything they can to make sure they continue to be employed and have interesting work. For them, an offer from an acquirer that commits to hire the employee base, perhaps paying them in line with industry standards after years of having been paid "startup wages," might be the best definition of success they can imagine.

And for others, the ability to walk away today with, let's say, $10 million in their pocket and the next decade open for them to try something completely new might be a much more successful outcome than continuing to navigate the stress of the Scale Wave and the risk of capsizing before finding that billion-dollar exit.

> You're an entrepreneur, not an investor. So think like an entrepreneur, not like an investor.

The only way to truly avoid the mistake of rejecting the right acquisition offer is by blocking out the noise of what everyone else wants from you and articulating, up front, what your own definition of success is.

Key Takeaways

---

1 Some acquisition offers should not be accepted. But more offers should not have been rejected.

---

2 Founders walk away from good acquisitions due to optimism bias. Your odds of raising money are highly correlated to your odds of getting acquired, making founders overconfident about getting through their next wave.

---

3 If you articulate, up front, what your own definition of success is, it'll help you know whether an acquisition offer is the right one for you.

Sidebar:
Debunking the M&A Critics

A March 2021 *Harvard Business Review* article by Clayton Christensen, et al. touts, "Companies spend more than $2 trillion on acquisitions every year. Yet study after study puts the failure rate of mergers and acquisitions somewhere between 70% and 90%."[19] This misses the point.

You can't consider the success rate of acquisitions in a vacuum. Rather, you must do so within the context of other options you have for growth. Talking about the success rate for M&As is like having half a conversation. The full conversation requires comparing the M&A success rate to the success rates of alternative growth strategies.

The business mantra is to decide whether to "build, buy, or partner." These are the three overarching categories for growing a business — build a product yourself (R&D), buy a business that's already built that product (M&A), or partner with someone to combine your resources to deliver the product to market.

When assessing the success of the money spent on M&A, why don't critics compare it against the success of money spent on R&D? Some of the most venerated companies have been shown to invest millions of dollars, even billions of dollars, on internal R&D, never to turn that investment into a profitable business. Meta Reality Labs division — focused mainly on its products in augmented reality (AR), virtual reality (VR), and the metaverse — has lost $45 billion over four years.[20] In a paper published by Microsoft employees, they write:[21]

"It is humbling to see how bad experts are at estimating the value of [product] features . . . Now that we have run many experiments, we can report that Microsoft is no different . . .

---

19 Christensen et al., "The Big Idea."
20 Khorram, "Meta's reality," Yahoo Finance.
21 Kohavi et al., "Online Experimentation."

only about one-third [of product features] were successful at
improving the key metric!"

OpenView, a premier venture capital firm, posted this on their blog in 2016: "Between 65% and 75% of new offerings fail outright or miss their revenue or profit goals, depending on whose research you look at. And the tab of those failures is hefty — $260 billion in the US alone in 2010."[22] Imagine what that tab looks like over a decade later.

The point isn't to scare you off investing any further in your R&D efforts. It's simply to add some perspective to the clickbait articles scattered across the internet about how bad a rate of return most M&As are. The reality is that growing a business is a risky and challenging endeavor, regardless of whether you're growing it organically through your internal R&D efforts, or inorganically through acquisition. Hopefully, by heeding our warnings about the biggest mistakes startups make in both of these areas, you can increase your odds of success.

---

22 Ramanujam, "This is Why 75% of New Products," OpenView.

Exit Wave Wrap Up

The potential for your business to get acquired is there whether you're still navigating your Launch Wave, struggling through the Pivot Wave, or riding the Scale Wave. Whether you develop that potential or let it waste away will determine your available options through your startup journey. The numbers show that most startups either fail to grow or fail altogether. But it doesn't have to be that way. We believe most startups do manage to develop something valuable. That value can be as specific as a team that has developed expertise in an area and learned to work together well. And it can be as broad as a scaled operating business with thousands of customers and multiple product lines. Both extremes, and everything in between, can deliver value to a larger acquiring company.

The biggest mistake founders make is thinking an acquisition is a transaction they might have to go through at some point in their journey, rather than a strategy they need to develop along the entirety of their voyage. Preparing for an exit is not a thing you start doing when you've decided you're tired of running the business, or you're running out of cash, or when your pivots aren't working. Starting an acquisition process then is a sure way to destroy the value you've worked so hard to build. By waiting too long to prepare for the Exit Wave, you end up desperately in search of a life raft before the wave sinks your ship, instead of being invited to join a fleet of world-class naval ships.

Throughout this section, we've explored how startups can unwittingly sabotage their acquisition prospects by: waiting until they're nearly out of money before thinking about an exit strategy, pitching the growth of their business instead of how their business can help the acquiring company grow, or turning down acquisitions when their confidence is at an all-time high — only to be hit hard by the next wave that they didn't prepare for.

The biggest mistake founders make is thinking an acquisition is a transaction they might have to go through at some point in their journey, rather than a strategy they need to develop along the entirety of their voyage.

To give yourself optionality and increase your chances of a successful outcome, it's critical that you consider your growth strategy in light of what your exit opportunities might be. This means taking an outside-in approach to your industry and investing in getting to know the larger companies within it. In particular, get to know their internal two-to-three year strategic plans — so you can first determine if your direction lines up with theirs and then make an informed decision about whether you might want to steer your ship alongside where they're going to better position yourself as an acquisition target. The "if you build it, they will come" approach rarely works when you're trying to get customers or trying to attract acquirers. And once you develop those key relationships with larger companies, increasing your optionality means understanding how to position your company to solve the key pain points driving the acquirer to consider buying its way through the pain.

The journey to acquisition is fraught with uncertainties and complexities. Not every startup that seeks acquisition will succeed, and not every acquisition will yield the anticipated benefits. But getting acquired represents a culmination of the hard work and perseverance you've put into your business and a validation of the innovation you created. But even though an acquisition comes last, the planning for it should be built into the structure of the company from its early days.

How You Can
Steer Your Startup
to Success

We've developed a culture that glorifies startup heroes as if they embarked on a flawless journey to success. It's exciting to read articles celebrating acquisitions that appear to have magically come together. Yet if you're lucky enough to hear the real stories from leaders of these companies — leaders like Darrell Benatar, Dave Garr, Bob Tinker, Laura Marino, Ashley McLain, and others we profiled in *Sail to Scale* — you'll see that these "overnight successes" were years in the making. And the few outliers that do magically come together create bad mental models for founders.

It's never fun to relive what went wrong, but it's by reflecting on how we navigated through the rough times that we learn and grow — both in business and our personal lives. It takes a certain amount of vulnerability to be able to do this. The idea for this book began at an offsite the three of us attend nearly every year. It's a place where executives have learned to share their mistakes, get feedback from one another in a safe space, and learn how to navigate their own approaching waves. It's a place where vulnerability is accepted, even celebrated — a rare experience in our world of social media-fueled perfectionism.

From these offsites, from our other networks, and from our research hearing from the many leaders we talked to in preparing *Sail to Scale*, we recognized a through line underlying how the best leaders managed to successfully navigate the mistakes often made during the Four Waves they faced as they grew their businesses.

Most importantly, we believe that becoming a successful entrepreneur is a learned skill — a learning journey — just like becoming a great software engineer or a great teacher.

The learning happens:

> Through research — like the research it takes to become a futurist, to see The Crack in the Market, to understand your customer's journey, and to understand your acquirer's pain points.

> Through incorporating feedback — whether you're listening to feedback on your MVP, from customers who aren't buying more of your product, or from your potential acquirers.

> Through doing — actually building something; actually trying to sell to someone; actually hiring, training, and sometimes firing; and actually doing everything that has to be done at least once before you pass it off to someone else to own.

> And finally, rinsing and repeating — researching, getting feedback on what you've done, and then doing it again, but better.

The learning journey helps you avoid the potentially fatal mistake of imposing your own beliefs and assumptions on a market that isn't interested.

We also believe building a successful business is not an individual sport. It's not a hero's journey. It's an endurance relay race of many heroic teams. It requires you to surround yourself with the right archetypes at the right times. It's a relay, because you'll most likely have to switch out your team to match the wave you're facing. And you'll have to manage egos along the way, including, and likely especially, your own. This is where building out your own advisory network becomes critical. *Sail to Scale* is rooted in stories from other founders because we believe that focusing only on our own experiences could distort the learnings through our lens. Drawing on the experiences of others has helped each of us swap out our own egos with the humility needed to deliver the insights we wanted to collect. We encourage you to find the network that works for you and gather even more learnings.

Finally, we believe the best entrepreneurs are the ones who know the wave they're best suited to navigate. Maria's secret sauce is discovering future trends and building teams that can deliver products that fit within those trends. Heather's superpower is seeing new opportunities in existing businesses and helping them harness those opportunities within the constraints of the business they've already built. Mona's X factor is driving the internal focus and the external relationships necessary to scale the business to new heights. Which wave best matches your special talents?

Entrepreneurship is not for the faint of heart. But being part of building something that others love is undoubtedly one of the most rewarding experiences you can have. The journey is not a straight line to success. And the true heroes of entrepreneurship aren't those who never speak of how they've faltered, but rather those who share their mistakes, so we can all rise stronger. Liberate yourself from the pressure of perfection. We hope you'll refer back to *Sail to Scale* as a guide as you approach each new wave along your voyage.

The world awaits the impact that only you can make.

Acknowledgments

As we reflect on the experience of writing *Sail to Scale*, we're not sure if it wasn't just a bit harder than the products we had to launch, the companies we had to turn around, and the companies we had to sell. Honestly, we would have benefited from a book laying out some of the biggest mistakes authors make when writing a book. A little irony for the road.

Just like launching a business, we couldn't have launched *Sail to Scale* without an incredible team of people, handpicked by us to ensure we had all the necessary archetypes covered.

Thank you to our publisher, Michelle Newcome and the How2Conquer team — Lauren Kelliher, Emily Owens, Charlotte Bleau, Telia Garner — for filling our operations archetype need and keeping us on track, making sure we didn't forget to write an introduction or, say, this acknowledgment, working tirelessly behind the scenes, and teaching us how complicated the world of publishing can be. Thank you to Karen Alexander, for being our conductor, making sure the parts we each wrote worked together, and telling us the hard truth when our words were more noise than music. Your invaluable feedback has been instrumental in shaping *Sail to Scale* into something we're immensely proud of. Thank you to Sandra Poirier Smith and the Smith Publicity team — Janet Shapiro, Olivia McCoy, Sophia Moriarty, Emily Willette — for filling our creative archetype, and formulating how we tell all of you about what we've written. Working with three authors is not easy! We couldn't have asked for a more dedicated and passionate group of professionals to work alongside this past year.

It wouldn't have been an interesting book without the real stories we could share with you from real entrepreneurs and business builders. We asked for your time, and you generously gave it to us so others could learn from your successes and mistakes. Thank you Darrell Benatar, Jennifer Coogan, Dave Garr, Dave Hansen, Dmitri Ignakov, Ilija Jovanovic, Laura Marino, Ashley McLain, Shannon Power, Jennifer Susinski, and Bob Tinker, for trusting us to tell your stories. Your courage and resilience serve as a beacon of inspiration for startups everywhere. Each of you has left an indelible mark on *Sail to Scale*.

A special thank you goes to HiPower and the incredible women in that network. HiPower is not only the place where the three of us met, but also the engine that runs across many of the opportunities we take, and this project is no exception, as we started it at a HiPower event. The executives of HiPower are an endless source of inspiration, and even more importantly, a selfless supply of practical advice, business connections, and a lot of the stories you read in *Sail to Scale*. This is a HiPower book.

We hope our work does justice to the contributions you've all made and that it resonates with entrepreneurs in the way it has been inspired by them.

Writing *Sail to Scale* while working full time has necessarily taken time from our family and friends. To our partners, children, parents, siblings, and friends who've patiently endured our late nights, early mornings, and endless chatter about startup woes and triumphs — thank you for your unwavering support through the years. Thank you to our loved ones, Guillaume, Eliott, and Oscar. Thank you, Jilly and Teddy. Thank you, Joe, Novak, Sascha, and Kira. We understand how hard it is to live with and love people like us, who are constantly embarking on new crazy projects. You put the wind in our sails.

And as we set sail for new adventures, from the bottom of our hearts, thank you.

— Mona, Heather, and Maria

Resources

Visit **h2c.ai/s2s** or use the QR code below for our recommended reading list, glossary, guided discussion questions, and other helpful resources from *Sail to Scale*.

Bibliography

Azhar, Ali. "Databricks Acquires Team Behind AI Startup Einblick." Databricks. Last modified January 31, 2024. https://www.datanami.com/2024/01/31/databricks-aquiresteam-behind-ai-startup-einblick/.

Bailyn, Evan. Average Customer Acquisition Cost (CAC) By Industry: B2B Edition. First Page Sage, 2024. https://firstpagesage.com/reports/average-customer-acquisition-cost-cac-by-industry-b2b-edition-fc/.

BBC. "Facebook and Zynga to end close relationship." November 30, 2012. https://www.bbc.com/news/technology-20554441.

Chartier, John, Alex Liu, Nikolaus Raberger, and Rui Silva. "Seven rules to crack the code on revenue synergies in M&A." McKinsey & Company. Last modified October 15, 2018. https://www.mckinsey.com/capabilities/growth-marketing-and-sales/our-insights/even-rules-to-crack-the-code-on-revenue-synergies-in-ma.

Chew, Louis. "Charlie Munger: The Power Of Not Making Stupid Decisions." Constant Renewal. Last modified July 17, 2018. https://constantrenewal.com/avoiding-stupidity/.

Christensen, Clayton M., Richard Alton, Curtis Rising, and Andrew Waldeck. "The Big Idea: The New M&A Playbook." *Harvard Business Review*, March 2011.

Coogan, John. "How Jason Citron Built Discord." John Coogan. Last modified June 14, 2021. https://www.johncoogan.com/how-jason-citron-built-discord/.

"Disagree and commit to ship things faster." *50 Folds* (blog). https://www.alexanderjarvis.com/disagree-and-commit-to-ship-things-faster/.

First Round. Episode 6, "Unpacking 5 of Atlassian's Most Unconventional Company-Building Moves." In *In Depth*. Podcast, audio, 96. https://review.firstround.com/podcast/unpacking-all-the-non-consensus-moves-in-atlassians-story-jay-simons/.

Friedman, Wayne. "Netflix Boasts Best Monthly Churn Rate, Disney+ Comes In Second." MediaPost, 14 Apr. 2021, www.mediapost.com/publications/article/362337/netflix-boasts-best-monthly-churn-rate-disney-co.html.

Gaspar, Carlos. "Operational Excellence." Milliken. Last modified November 5, 2023. https://www.milliken.com/en-us/businesses/performance-solutions-by-milliken/blogs/importance-of-an-operational-excellence-based-culture.

Graham, Molly. "'Give Away Your Legos' and Other Commandments for Scaling Startups." Interview. *The First Round Review* (blog). https://review.firstround.com/give-away-your-legos-and-other-commandments-for-scaling-startups/.

Gustin, Sam. "Groupon Fires CEO Andrew Mason: The Rise and Fall of Tech's Enfant Terrible." *Time*, March 1, 2013. https://business.time.com/2013/03/01/groupon-fires-ceo-andrew-mason-the-rise-and-fall-of-techs-enfant-terrible.

Haden, Jeff. "Amazon Founder Jeff Bezos: This Is How Successful People Make Such Smart Decisions." *Inc.*, December 3, 2018. https://www.inc.com/jeff-haden/amazon-founder-jeff-bezos-this-is-how-successful-people-make-such-smart-decisions.html.

"How to drive innovation through ecosystems and partnerships." EY. Last modified March 6, 2020. https://www.ey.com/en_nz/consulting/how-to-drive-innovation-through-ecosystems-and-partnerships.

Huang, Jensen. Interview. *NVIDIA CEO Jensen Huang*. Produced by Acquired. 2023. https://www.acquired.fm/episodes/jensen-huang.

Huston, Caitlin. "Take-Two Completes $12.7B Acquisition of Zynga." *The Hollywood Reporter*, May 23, 2022. https://www.hollywoodreporter.com/business/business-news/take-two-zynga-acquisition-1235152410/.

"Intuit to Acquire Mailchimp." *Business Wire*, September 13, 2021. https://www.businesswire.com/news/home/20210913005806/en/.

K., Induprakas. "'Repetition doesn't ruin the Prayer' and nine other things I learned from Brad Smith at Intuit." LinkedIn. Last modified February 12, 2020. https://www.linkedin.com/in/induprakas?trk=article-ssr-frontend-pulse_publisher-author-card.

Kohavi, Ronny, Thomas Crook, Roger Longbotham, Brian Frasca, Randy Henne, Jean Lavista Ferres, and Tamir Melamed. "Online Experimentation at Microsoft." Microsoft Think Week paper, 2009. PDF.

Kotter International. Last modified 2024. https://www.kotterinc.com/.

Lyons, Dan. "Groupon IPO No Bargain as Early Investors Milk the Company." *The Daily Beast*, July 13, 2017. https://www.thedailybeast.com/groupon-ipo-no-bargain-as-early-investors-milk-the-company.

Mac, Ryan. "Groupon Shares Crumble After Company Names New CEO." *Forbes*, November 3, 2015. https://www.forbes.com/sites/ryanmac/2015/11/03/groupon-names-coo-rich-williams-as-new-ceo-former-chief-eric-lefkosky-becomes-chairman/.

McGrath, Maggie. "Coca-Cola Buys Stake In Monster Beverage For $2 Billion." *Forbes*, August 14, 2014. https://www.forbes.com/sites/maggiemcgrath/2014/08/14/coca-cola-buys-stake-in-monster-beverage-for-2-billion/?sh=4eb553175427.

Microsoft News Center. "Microsoft to acquire Activision Blizzard to bring the joy and community of gaming to everyone, across every device." Microsoft. Last modified January 18, 2022. https://news.microsoft.com/2022/01/18/microsoft-to-acquire-activision-blizzard-to-bring-the-joy-and-community-of-gaming-to-everyone-across-every-device/.

Murphy, Lincoln. "What is a good SaaS Churn Rate?" Customer-centric Growth by Lincoln Murphy. Last modified 2013. https://sixteenventures.com/saas-churn-rate.

O'Brien, Matt, and Haleluya Hadero. "OpenAI brings back Sam Altman as CEO just days after his firing unleashed chaos." *Associated Press*, November 22, 2023. https://apnews.com/article/altman-openai-chatgpt-31187f7f6eca8ff9d0eef7585aac6ace.

O'Reilly Media, Inc. "The Laws of Learning." O'Rielly. Last modified 2024. https://www.oreilly.com/library/view/wooden-a-lifetime/9780071507479/ch172.html.

O'Shaughnessy, Patrick. Episode 330, "Henry Schuck - Building ZoomInfo." May 23, 2023. In *Invest Like the Best*. Podcast, audio, 92. https://www.joincolossus.com/episodes/49638569/schuck-finding-your-next-best-customer?tab=transcript.

Pitchbook. *PitchBook-NVCA Venture Monitor Q3 2023*. PitchBook-NVCA, 2023. https://pitchbook.com/news/reports/q3-2023-pitchbook-nvca-venture-monitor.

Poschenrieder, Martin. "What is an Amazon Six Pager Memo?" *Six Pager Memo* (blog). Entry posted July 22, 2023. https://www.sixpagermemo.com/blog/what-is-an-amazon-six-pager.

Ramirez, Olivia. "How Demandbase Acquired DemandMatrix in 7 Months After Launching a Partnership." *ELG Insider* (blog). Entry posted August 31, 2021. https://insider.crossbeam.com/resources/how-demandbase-acquired-demandmatrix-after-partnership.

Ramanujam, Madhavan. "This is Why 75% of New Products Fail." OpenView. Last modified July 19, 2016. https://openviewpartners.com/blog/why-new-products-fail.

Reichheld, Fred. "Prescription for Cutting Costs." Bain & Company. Last modified October 25, 2001. https://www.bain.com/insights/prescription-for-cutting-costs-bain-brief/.

Rogers, Everett M. "Adopter categorization on the basis of innovativeness." Chart. In *Diffusion of Innovations*, 3rd ed., by Everett M. Rogers, 247. New York, NY: The Free Press, 1962.

Salesforce. "New Study Finds Salesforce Economy Will Create 9.3 Million Jobs and $1.6 Trillion in New Business Revenues by 2026." Salesforce. Last modified September 20, 2021. https://www.salesforce.com/news/press-releases/2021/09/20/idc-salesforce-economy-2021/.

Sanctuary, Hillary. "Where to play: a practical guide for running your tech business." EPFL. Last modified September 26, 2017. https://actu.epfl.ch/news/where-to-play-a-practical-guide-for-running-your-t/.

Satir, Virginia. *The Satir Model: Family Therapy and Beyond*. United States: Science and Behavior Books, 1991.

Scope AR. Last modified 2024. https://www.scopear.com/.

Seuss, Dr. *Oh, The Places You'll Go!* New York, NY: Random House Children's Books, 2013.

Soltys, Douglas. "Xero Acquires Toronto's Hubdoc for $70 Million USD." *BetaKit*, July 31, 2018. https://betakit.com/xero-acquires-torontos-hubdoc-for-70-million-usd/.

Teller, Astro. "Google X Head on Moonshots: 10X Is Easier Than 10 Percent." *Wired*, February 11, 2013. https://www.wired.com/2013/02/moonshots-matter-heres-how-to-make-them-happen/.

Vogel, Marco. "Broadcom's acquisition of VMware: What you need to know." SoftwareOne. Last modified March 25, 2024. https://www.softwareone.com/en/blog/articles/2024/03/25/broadcoms-acquisition-of-vmware-what-you-need-to-know.

Weinberg, Cory. "Software Startup That Rejected Buyout From Microsoft Shuts Down, Sells Assets to Nutanix." The Information. Last modified December 8, 2023. https://www.theinformation.com/articles/a16z-backed-startup-that-once-rejected-150m-sale-to-microsoft-shuts-down.

"Why startups failed in 2022." *Skynova* (blog). Entry posted 2023. https://www.skynova.com/blog/top-reasons-startups-fail.

Wollenweber, Kevin. "Cisco Announces Intent to Acquire Accedian: Accelerating High Performance Service Assurance." *Cisco* (blog). Entry posted September 29, 2023. https://blogs.cisco.com/news/cisco-announces-intent-to-acquire-accedian.

Meet the Authors

Mona Sabet has been called an indomitable force in scaling both businesses and communities. She has spent a career spanning over 25 years at the vanguard of the technology industry, steering numerous technology enterprises from their launch stage to exits, either by acquisitions or public offerings. Mona's expertise is focused on growing businesses through a combination of relentless focus on scaling business operations, building global partnerships, and driving strategic acquisitions.

Mona has held a number of executive roles at B2B technology companies including as Chief Corporate Development and Administration Officer at VulcanForms, Chief Corporate Strategy Officer at UserTesting and Corporate Vice President at Cadence Design Systems (NASDAQ: CDNS).

At VulcanForms, she orchestrates alignment among business operations, commercial activities and overarching financial objectives. Her leadership at UserTesting culminated in the company's IPO on the New York Stock Exchange. Mona's career also spans M&A roles, both at Cadence Design Systems and as Managing Partner at Tribal Advisors and Corporate Vice President.

Beyond the C-suite, Mona serves on boards and advisory boards, championing the cause of startups and nurturing the ecosystems that sustain them. In addition to her corporate leadership, Mona co-founded ChIPsNetwork.org, a global community for women in tech, law, and policy. She also founded and continues to lead HiPower, a vibrant community of leaders championing executive women poised to shape the future of technology and beyond.

Heather Jerrehian is an entrepreneurial futurist and investor who has driven innovation and go-to-market strategies for startups and Fortune 500 companies for over two decades.

Possessing an extraordinary ability to foresee market movements and consumer trends, Heather has consistently led cutting-edge enterprises toward significant success. As a serial entrepreneur and technology executive, Heather has built and scaled multi-million dollar ventures, leading turnarounds at unprecedented speed, achieving 10X growth, and engineering profitable exits, including the acquisition of Hitch by ServiceNow (NYSE: NOW) under her leadership as CEO.

At ServiceNow, she works at the forefront of social and technology shifts, building next-generation AI solutions and workforce intelligence products as the VP of Product Management for Employee Workflows. As COO at Emtrain, her leadership was instrumental in doubling the HR Tech company's revenue and turning multi-million dollar losses into profits within 18 months. Heather's vision and foresight led to a new $25M division at Drop, where she launched over 100 proprietary products across various categories, from hardware and electronics to clothing and accessories.

Heather is a Founding Limited Partner of How Women Invest, a venture capital firm aimed at increasing investments in female-founded companies, and serves on Fast Company's Executive Board. A sought-after speaker, Heather was recognized as a Woman of Influence by the Silicon Valley Business Journal.

Maria Fernandez Guajardo is a transformative product executive and entrepreneur with an extensive history of innovation in the tech industry. Maria has guided companies of all sizes, from startups to giants like Meta and Google, in bringing cutting-edge products for consumer and B2B to market at a global scale.

Presently, Maria serves as a Senior Director of Product Management at Google, leading the product team at Gmail, enhancing user productivity with AI innovations. As a product executive at Meta, Maria oversaw the transformation of Facebook's video and audio consumer offerings. There she also funded and led "VR for Work" metaverse applications which brought Oculus virtual reality into the productivity and corporate realm.

Before that, Maria held the position of Vice President of Product at Clear Labs and RetailNext, focusing on food genomics and pioneering computer vision-based retail analytics, respectively. Maria holds a Bachelor of Science in Electrical Engineering and began her career as a silicon engineer at Cadence Design Systems followed by holding varied engineering and business roles at Texas Instruments in France and the US.

Maria is committed to promoting diversity through her non-profit engagements. She founded the Silicon Valley Club de Ejecutivas Españolas and served on the board of STEM non-profit Ignited Education. Maria has been recognized as one of the top 40 leaders in the Bay Area by the *Silicon Valley Business Journal* and as a top Latino influence in tech by *CNET*.

Mona (left), Maria (middle), and Heather (right)

Thank You for Joining Us

Dear Reader,

It is with our heartfelt gratitude that we thank you for accompanying us on this journey across our Four Waves of startup growth. Our hope is that you've found the ideas presented here not only informative but also transformative.

We understand that your time is valuable, and we sincerely appreciate you choosing to spend it with our book. Your engagement and commitment to learning are what drive authors like us to continue sharing our experiences and insights.

Share Your Thoughts

If you've found this book valuable and believe it can benefit others, we encourage you to share your thoughts. Reviews play a crucial role in helping more readers discover our work on platforms like Amazon. Your honest review can make a significant difference in reaching those who could benefit from the ideas discussed here.

Here's how you can write a review on Amazon:

1. Visit the Amazon product page: Search for *Sail to Scale* on **amazon.com.**

2. Log in to your Amazon account: If you haven't already logged in, you'll need to do so to leave a review.

3. Click on 'Write a Customer Review': Scroll down to the Customer Reviews section of the book's page. You'll find a button that says 'Write a customer review.' Click on it.

4. Rate the book: Choose a star rating that reflects your overall opinion of the book. Then, proceed to write your review.

5. Write your review: Share your thoughts on what you liked about the book, how it has impacted your understanding, and any insights you gained. Your review doesn't need to be long — just genuine and helpful to other potential readers.

6. Submit your review: Once you're satisfied with your review, click 'Submit.' Your review will then be published on the book's Amazon page.

Join Our Community

We invite you to stay connected with us and join our growing community of readers. Visit us at **sailtoscale.com** and sign up to join our mailing list. We'll keep you updated on additional resources, new releases, exclusive content, speaking engagements, and more.

Thank you in advance, and we wish you smooth sailing!

Printed in the USA
CPSIA information can be obtained
at www.ICGtesting.com
JSHW012130260924
70377JS00004B/7/J